USER-CENTERED TECHNOLOGY

SUNY series, Studies in Scientific and Technical Communication
James P. Zappen, Editor

USER-CENTERED TECHNOLOGY

A Rhetorical Theory for Computers and Other Mundane Artifacts

Robert R. Johnson

STATE UNIVERSITY OF NEW YORK PRESS

Published by
State University of New York Press, Albany

Printed in the United States of America

For information, address State University of New York Press,
State University Plaza, Albany, N.Y., 12246

Production by Cathleen Collins
Marketing by Anne Valentine

Library of Congress Cataloging in Publication Data

Johnson, Robert R., 1951 –
User-centered technology : a rhetorical theory for computers and other mundane artifacts / Robert R. Johnson.
p. cm. — (SUNY series, Studies in scientific and technical communication)
Includes bibliographical references and index.
ISBN 0-7914-3931-3 (alk. paper). — ISBN 0-7914-3932-1 (pbk. : alk. paper)
1. User interfaces (Computer systems) 2. Human engineering.
3. Technology—Social aspects. I. Title. II. Series
QA76.9.U83J64 1998
808′.0666—dc21 97-52146
CIP

10 9 8 7 6 5 4 3 2 1

Contents

Figures

Foreword

The State University of New York (SUNY) Press's series in Scientific and Technical Communication seeks to broaden our understanding and improve our practice of scientific, technical, and professional communication by publishing scholarship and research from a variety of disciplines directed toward these ends. During the past twenty years, scholarship and research in scientific and technical communication have become increasingly broad in scope and increasingly sophisticated, drawing upon such diverse disciplines as cognitive psychology and linguistics, the history and philosophy of science and technology, ethnography-based composition studies, rhetoric and literary theory, and, more recently, intercultural communication, cultural studies, critical theory, human–computer interaction, and visual design, among many others. The rapid development of Internet-based communication practices promises to extend and complicate this research even more.

This growing body of scholarship and research has, we must hope, greatly improved our understanding and practice of scientific and technical communication, but it also has become increasingly divided and conflicted. Researchers from various disciplines have differed about the relative merits of their respective research agendas, and practitioners and teachers of technical communication have found some of these agendas more useful than others. For example, proponents of readability and accessibility research, usually based in cognitive psychology and linguistics, believe they can identify, and have identified, sound and widely applicable principles of effective communication. Ethnographers observing differences in organizational cultures believe that what counts as effective communication practice differs within various organizational cultures. Rhetoricians seek to serve both organizational and social goals, technical writers and their organizations, and also the larger society, the end users of technology. Cultural and critical theorists critique what they perceive to be a narrow organizational and instrumental orientation in all of this research. Such differences typify the current state of scholarship and research in scientific and technical communication.

More research will not necessarily resolve these differences, which are rooted in our most deeply held values and beliefs about the world and our relationships with each other. But it can help us test these values and beliefs against the standards of increased understanding and improved practice and pedagogy. Robert Johnson's contribution to this SUNY Press series seeks to improve our understanding, practice, and pedagogy by offering a rhetoric-based theory of user-centered practice. As theory, it draws upon the ancient notion of rhetoric as an art, or a *technē*, traditionally the art of persuasion, but more recently and more broadly the art of making meaning and motivating action in human situations of all kinds. As practice and as pedagogy, it seeks to shift the emphasis in technical documentation from the user as a passive recipient of technical information to the user as an active cocreator of that information and the technology it inscribes.

As the Internet and other computer-based technologies make the world smaller, and as cultural differences become closer and more apparent to us, such a strong user-centered understanding of technology and technical communication will come to seem both more humane and more practical. If the SUNY Press's Studies in Scientific and Technical Communication can contribute to a better understanding and an improved use of technology and technical information in such a global setting, then it will have done its work.

James P. Zappen
Rensselaer Polytechnic Institute

Preface

Many scholarly fields associated with technology have attempted to ask, in one way or another, a fundamental question: What is the relationship between humans and technology? To address this question, each discipline has brought its own special "twist" to the complex of answers that surround this question. In philosophy and political theory, Langdon Winner pursued this question in *Autonomous Technology: Technics Out-of-Control as a Theme in Political Thought,* arguing that it is imperative for modern society to begin making decisions about how we should take action to counter the powerful, but ultimately controllable, force of technology. He argues that we have collected more than enough information about technology and its consequences and that we now should turn our attention to setting a course that will allow us to make useful changes in how we live with technology.

> What we lack is our bearings. The contemporary experience of things technological has repeatedly confounded our vision, our expectations, and our capacity to make intelligent judgments. Categories, arguments, conclusions, and choices that would have been entirely obvious in earlier times are obvious no longer. Patterns of perceptive thinking that were entirely reliable in the past now lead us systematically astray. Many of our standard conceptions of technology reveal a distortion that borders on dissociation from reality. And as long as we lack the ability to make our situation intelligible, all of the "data" in the world will make no difference. (pp. 7–8)

Ultimately, Winner presents an approach to technology based upon concepts of *limits*: limits that we have somehow lost sight of in the present century as technological innovation and deterministic attitudes of progress endanger our ability to make intelligent, situated decisions regarding technological development. The act of defining these limits and then governing them is, in his view, the responsibility of the populace, not something that should just be left to the institutions that currently control the direction of technology design, development, and dissemination.

Another philosopher, Carl Mitcham, approaches technology as a thinker. That is, his recent book—*Thinking Through Technology: The Path Between Engineering and Philosophy*—stresses the need in modern times to ponder what technology is, how it works, and what the parameters are of its moral and ethical dimensions. His contribution to the study of technology (at least part of his contribution) is the forging of a pathway into human consciousness about technology's role in our day-to-day lives.

> For many people . . . the ideas that guide their lives may not be held with conscious awareness or full articulation. They often take the form of myth. Philosophical argument and discussion introduce into such a world of ideas a break or rupture with the immediately given. This rupture need not require rejecting or abandoning that given, but it will entail bringing the given into fuller consciousness or awareness, from which it must be accepted (or rejected) in a new way or on new grounds. (p. 277)

Mitcham wants the discipline of philosophy to create a space where people can become critically aware of technology and its manifestations. To this end, he presents a taxonomy of technology that bridges the historical and theoretical aspects of the applied side of technology (which he terms the "engineering philosophy of technology") and the intellectual side (which he dubs the "humanities philosophy of technology").

Critical and cultural theorists have also turned their gaze upon technology, notably in the book by Andrew Feenberg, *Critical Theory of Technology*. Feenberg calls for a view of technology that eventually would redefine " . . . the paradigm of technical practice . . . " through " . . . a return on a rational basis to the recognition of the natural and human constraints on technical development" (p. 197). Similar to Winner, but clearly more couched in a Marxist, socialist frame, Feenberg brings the disciplines associated with critical theory into the debates surrounding technology, technique, and society.

Sociologists have embraced the issues of technology and humans in fairly large numbers in recent years, turning out several important collections. Maybe most important, they have helped develop methods of investigation (mostly ethnographic or case studies) that tell stories of people and technology in a variety of contexts. They also have combined efforts with historians of technology to refine these methods and, in turn, have added the historians' expertise with textual study to broaden the breadth of their investigations.

Feminist scholars have most recently come to the stage of the technology debates and have put extremely interesting and positive twists on the human/technology issues. In a fine overview of feminist perspectives, Judith Wacjman (1991) reminds researchers of technology and culture that we should be interested in the private spheres: spheres that can illuminate the concrete "lived experience" of women and other marginilized groups. She argues for a move (albeit carefully) away from the abstract theorizing of technological phenomena toward a more detailed analysis of concrete situations: "The character of salient interests and social groups will differ depending on the particular empirical sites of technology being considered. Thus we need to look in more concrete and historical detail at how, in specific areas of work and personal life, gender relations affect the technological enterprise" (p. 24). She further complicates the issue of gender itself by pointing out that it is "... [o]ften difficult to disentangle the effects of gender from those of class and race" (p. 24). Thus, the study of technology and human endeavor becomes more problematic—and more interesting—as different disciplines enter the fray.

I could continue this review for quite some time. There are, for instance, historians of technology who have blazed some of the earliest trails through the fascinating study of technology and culture. But we will wait to meet many of them in later chapters. For now, I want to raise the question, where are the technical communicators in this important field of scholarship? Without a doubt we have forged some paths into this murky but rich swamp of ideas. Much of our attention to the relationship between people and technology has been based on ways to redirect our approach to the technical communication classroom. These are worthwhile endeavors, to say the least, and in fact rethinking pedagogy is something that I am sure we do much better than the fields I have previously mentioned. We could move into technology discussions from this perspective, and we should (more on this later). We should, however, move more aggressively into the foray concerning technology and humans in a broader, more theoretically and historically based manner. We have had some success in this direction. Carolyn Miller, Charles Bazerman, Greg Myers, Dale Sullivan, and a few others have crossed these bridges, and the promise of such scholarship is exciting. I offer this book as a modest contribution to the effort of discussing technology critically within the frame of technical communication.

Defining the audience of this book has been an ongoing challenge. The primary audience (as the title of this State University of New York Press series makes clear) is technical communicators. My graduate training in rhetoric and composition, coupled with my specialization in technical communication, however, has given me an "academic split personality" that potentially complicates any straightforward definition of my audience. Consequently, this book has several audiences that I address one by one.

Clearly, I hope to contribute to scholarly research in technical communication. Technical communication is a young field, and as such it is only beginning to define its research methodologies, theoretical underpinnings, and historical bases. This book is a small contribution to the effort to build the disciplinary features of technical communication as an academic and a professional discipline. Many of the members of this research audience are, of course, also teachers of technical communication. There are a number of pedagogical examples that make the book relevant to those involved with the training and education of technical communicators.

For workplace technical communicators, this book is quite different from "how-to" or practical texts in that I am not laying out a systematic agenda for action. At the same time, there is much a writer can "take to work" from the pages of this book, especially concerning conceptual understandings of what writers do on a day-to-day basis. During my work over the past ten years in preparing technical communicators for the workplace, it has become clear to me that the conceptual, theoretical understanding of the "how" and "why" of technical communication practice is of ultimate importance to long-term success. Once technical communicators have acquired rudimentary skills, it is important that they have a deeper, more conscious understanding of what it is they do. This is one purpose of the book: to reveal some of the "givens" that we take for granted concerning users and technology as we go about our everyday lives writing manuals and reports, deciding what should be on-line and what should be in print, defining user audiences, and negotiating with managers over design and development decisions.

The second major audience consists of researchers and teachers of rhetoric and composition. For the researchers, my intent is to offer some perspectives on rhetoric that may, at times, surprise those whose scholarly interests have resided outside the boundaries of technical communication specifically, or technology studies generally. Issues pertaining to the arts of rhetoric, the problem of epistemology, and the relationship of rhetoric to technology could provide some insights for those involved in mainstream rhetoric and composition research.

I should add that I approach rhetoric studies unabashedly as a rhetorician who takes seriously the historical breadth of the discipline. I perceive rhetoric as a discipline that, for over twenty-five hundred years, has had a central investment in revealing the unconscious and uncovering the mysterious for the end of transferring knowledge in a democratic and an ethical manner. There have been, throughout history, those who have argued against this view of rhetoric. They fear rhetoric as the deceitful use of language for selfish ends, or as a hyperrational approach to the use of language based upon algorithmic techniques. I stand far from such perjorative views of rhetoric, and thus the arguments in this book do so as well.

Concerning the teacher of mainstream composition, the contents of this book, although somewhat afield of most of your usual concerns, will possibly provide a different view of what it means to be a writer. For instance, technical communicators pursue a variety of interests that might prove interesting to anyone who has not become involved with this form of writing. The following critiques of technology from a user's perspective are another view of modern culture to present to students in writing classes. I would caution, however, those who might consider the user-centered views in this book as a means of empowering the "subject"—the student writer. In fact, this book argues that the "writers" of technology, those who design and build technological artifacts and systems, are already empowered. Instead, this book works toward the empowerment of users: the audiences of technology. In plainer terms, I am arguing for an audience-centered, not a writer-centered approach to technology.

Finally, there is a third audience I cannot claim as primary, but nevertheless I would like to count among the readers. This audience is comprised of those scholars from the various disciplines that constitute the general field of technology studies. I borrow quite heavily from them to frame my arguments and provide examples of technology and use. I hope I do them justice in my borrowing, and I in turn hope they might find my writings to be of some interest.

The Structure of the Book

The book's seven chapters are placed into into three overall sections. Part 1—Situating Technology—consists of two chapters that describe the everyday contexts and theory of technological use. Chapter 1 introduces the concept of "mundane technologies" through a series of scenarios that depict the breadth of technology in our everyday lives. The second chapter presents the model of the "user-centered rhetorical complex of technology" and develops a rationale for it. The "complex" is my theoretical invention that attempts to situate the "end" of technology in users. The chapter also connects ancient and modern rhetoric within the context of technology and technical communication—a connection that serves as a thread throughout the book.

Part 2—Complicating Technology—is the theoretical and historical core of the book and consists of three chapters. In these chapters I work interdisciplinarily to lay a foundation for a user-centered approach to technology—an approach relevant to technical communicators, but one I hope is also of interest to scholars and teachers in rhetoric, composition studies, and related disciplines. Working from three central elements of the "complex" from chapter 2—*users, artisans/designers, and artifacts/systems*—I focus in each of the chapters on one of these elements. Chapter 3 looks at the "rhetorical element" of users, and argues for what I call "user knowledge." Drawing upon the history of rhetoric, and to some extent the history and philosophy of technology, I present a case for revising our sense of what users know about technology. In a nutshell, this chapter calls for a revaluing of users, and that such a revaluing can be accomplished through a better sense of what they know and bring to technologies.

Chapter 4 concentrates on the *artisan/designer* perspective of technology. Here I review the history of human factors research to make the case that although human factors research appears to be interested in users, in fact it is a discipline that most often is aimed at other concerns—primarily system efficiency for economic ends. I conclude that some human factors research is quite useful to technical communicators, but that we must be careful how we draw upon this research as we use it toward user-centered goals. Chapter 5 is an examination of the *artifact/system* perspective of technology. Here I investigate the issue of technological determinism from an interdisciplinary perspective that uses the fields of sociology, history, and philosophy as its base.

Part 3—Communicating Technology—is the final section of the book and consists of two chapters that are "applications" of the theoretical and historical bases of the first two sections. Chapter 6 applies the model of the user-centered rhetorical complex to technical communicators in the *nonacademic* sphere. The content of this chapter is a discussion of the writing of computer user documentation (both print and on-line forms) that is informed by the user-centered theory presented in the earlier chapters. I revise the model of the rhetorical complex to explicitly "fit" the practices of developing computer documentation. Chapter 7 concludes the book with two applications of user-centered concepts in the academic sphere. The first example is pedagogical—I use a case study of my own experiences teaching technical communication through a user-centered perspective. The second examines how academic technical communicators can apply user-centered concepts to historical research.

Acknowledgments

As a piece of work that began as graduate research, became a dissertation, expanded once again as the concept for a book, and then was continually revised as students and colleagues read drafts of various chapters, I have many people to thank.

From my graduate study at Purdue University I am indebted to Janice Lauer and Irwin Weiser. Their encouragement and advisement kept me on track with sage advice. I especially thank Pat Sullivan, who not only directed my committee but introduced me to the teaching and theories of technical communication. Nancy Allen, Jennie Dautermann, and Mark Simpson, three of my peers at Purdue, likewise provided much assistance and counsel both during and after graduate study.

As my earlier work has matured into a book, the list of those who have offered their help has grown exponentially. At Miami University, my technical communication colleagues Paul Anderson, Jennie Dautermann, Jean Lutz, and Gil Storms helped me from the early stages of the book proposal to the final drafts of the chapters.

In addition, Miami University's rhetoric and composition graduate program and its master's program in technical and scientific communication have afforded me the privilege of working with some of the finest students one could imagine. Many doctoral students have kept me on my toes, especially Francie Ranney, who makes sure I keep digging a little deeper into classical rhetoric, Ann Brady Aschauer, who reminds me continually of the connections between technical communication and feminist studies, and Jim Dubinsky, who keeps me thinking in interdisciplinary modes. Parag Budhecha, Serena Hansen, Dominic Micer, Malea Powell, and Pegeen Reichert Powell also have introduced me to many theories and methods, without which I would be much the poorer.

The master's students have helped me continually stay in touch with workplace realities through their client projects and internship experiences. Of this large group I would like to especially thank Gail Bartlett, Scott Deloach, Rob Houser, Natalie Larsen, Tim Kiernan, and Frank Sullivan for their assistance and support—practical, intellectual, and otherwise. Many examples in this book come from the experiences I have had with these "real world" technical communicators.

In the greater field of technical and professional communication, I have been unusually fortunate to receive support and encouragement from a variety of people. Jim Zappen has been most helpful throughout the entire development of this project; Jimmie Killingsworth encouraged me to do it; Barbara Mirel has been an incomparable friend and a supporter; and Steve Doheny-Farina has provided me with helpful suggestions at several stages. I want also to thank Greg Clark and his technical communication class at Brigham Young University for their honest and helpful feedback on several early drafts of chapters.

I owe a significant debt to Leta Roberson, the graduate secretary in the Miami English department, for making my day-to-day tasks of directing a graduate program easier and less stressful. I also want to thank Miami University for an Assigned Research Appointment that enabled me to complete several chapters of the book.

Finally (and of course not least), I am grateful to my family. My father, my sisters Judy and Carol, and my brother Tom have been supporters in more ways than one. To my children—Rachel, Tyler, Abby, and Drew—I say "Thank You" for putting up with several moves and often having little time to do things together. To Evie, well, I cannot say enough, except maybe, "It's done!"

PART I

Situating Technology

CHAPTER 1

Users, Technology, and the Complex(ity) of the Mundane

Some "Out of the Ordinary" Thoughts

> . . . the user, or, in other words, the master, of the house will even be a better judge than the builder, just as the pilot will judge better of the rudder than the carpenter, and the guest will judge better of the feast than the cook.
>
> —Aristotle, *Politica*

The word "mundane" is an adjective that, when converted to a noun, describes the essence of the world from a user point of view. According to the dictionary,[1] *mundane* means "common, ordinary, or, of this world." Users of technology work, play, and in virtually every way *live* in the world of the mundane—a complex and colorful realm that, ironically, remains practically invisible. The invisibility of the mundane is, I suppose, not surprising. As we go about the activities of our daily lives, doing things repeatedly each day, we internalize these actions and thus make them a part of our unconscious.

In part due to its invisibility, this site, this "land of the mundane," has not been the focus of widespread research, especially concerning the role that users of technology play in this land. Many researchers have focused on the designers and developers of technologies, providing us with important interpretations of how artifacts "come to be." Others have likewise

1. The *Random House Webster's Dictionary*, 1993.

concentrated their efforts on the artifacts and systems that comprise the technologies of human endeavor. In this book, we will instead turn our attention to the everyday users of technology. We will make room for the developers, the designers, and the artifacts or systems of technology, but the primary concern will be to examine users and the phenomena of technological use from their perspective.

In this opening chapter, I will lay some groundwork concerning issues relevant to users and technology that will be developed throughout this book, but the structure of this initial chapter will not be conventional. That is, it will not be a review of the literature per se, but rather it will raise issues and ask questions that are investigated later in the text. We will begin by rethinking what it means to be located in the mundane for the purpose of describing the complexity, and the resultant complex of interactions that evolve as one probes this relatively unexplored land. Next, I will tell some tales of my own experiences that will help uncover more concretely some of the issues and questions concerning the study of users, technology, and the mundane. Finally, we will ask what it means to develop a theory of the mundane through user-centered concepts.

Retelling and Reinventing the Mundane

Technology has long been the source of interesting stories. Many of these stories are full of glory and conquest—they are cast in the light of progress as it aims its beam into the dark, infinite beyond. We all are familiar with stories such as those, for instance, of Thomas Edison, the lone genius who worked feverishly night and day to perfect the lightbulb, the phonograph, and even the motion picture camera. We also know well tales of technological achievements such as the story of the American railroad, which was built piece by piece over mountains and through deserts, always moving positively and progressively toward the "West." This, of course, is the same railroad technology we now describe sentimentally, like an old, trusty pet: the dying iron horse whose day is past and who is doomed to a slow but comfortable death. Such are the tales we tell again and again of technology and its heroes.

But there are other stories to tell of technology, too. Hidden from view, almost imperceptible because they blend so perfectly into the backdrop of daily, mundane experience, are stories that beg to be told of people as they work with, against, and through the technologies that abound in our lives. These silent, hidden stories have been effaced in modern times, however, as the value placed upon the stories of everyday knowledge—of "know-how"—has given way to the "knowledge in the machine," or the "knowledge in the system." As Michel de Certeau puts it,

> . . . know-how takes on the appearance of an "intuitive" or "reflex" [sic] ability, which is almost invisible and whose status remains unrecognized. The technical optimization of the nineteenth century, by drawing from the reservoir of the "arts" and "crafts" [sic], the models, pretexts or limits of its mechanical inventions, left to everyday practices only a space without means or products of its own; the optimization constitutes that space as a folkloric region or rather as an overly silent land, still without a verbal discourse and henceforth deprived of its *manouvrier* [sic] language as well. (p. 69)

The knowledge of everyday practice has become nearly voiceless: a colonized knowledge ruled by the technology and the "experts" who have developed the technologies. Yet the voices and the knowledge embedded in the stories are still there—they are just more difficult to hear, more difficult to recognize. For de Certeau, *know-how* has become a matter of folklore, of tales often told but not believed to be "real." Thus, as folklore, the appreciation of *know-how* and of *use* has been lost because the *arts* of *know-how* that were at one time conscious have come to reside in the collective unconscious: not seen, not heard, and not known, a type of knowledge that has been stripped of its ability to consciously voice its purpose, power, and means by which it can make its knowledge visible. Rhetoric, particularly the *arts* of rhetoric, can be used to resurrect this lost form of knowledge and make it visible. In the coming pages of this book, we will consistently turn to the ancient concept of rhetorical arts to uncover, and revalue, the knowledge of know-how and use.

There also is a richness in these tales of the mundane that we too easily overlook: a richness that lifts the everyday experiences that we merely think of as *know-how* or *doing* to another level that transforms the mere *doing* into *living*. At this heightened level, there is a reciprocity among the artifact itself, the knowledge of know-how, and the social milieu that always circumscribes technology, knowledge, and actions. A potent example of this interchange is found in Helen and Scott Nearings' *The Maple Sugar Book*—a book that in the very best of ways can be termed a treatise on know-how or use. In the conclusion of their book, the Nearings reflect upon their years of learning about, and then actually farming, a self-sufficient maple sugar operation in southern Vermont:

> We have earned from maple and found a means of livelihood. We have also learned from maple. The occupation of sugaring has been a thorough-going education and broadened our contacts with life in its many aspects. The young Thoreau in his journal wrote, "Had a dispute with my father about the *use* [sic] of my

> making this sugar. . . . He said it took me from my studies. I said I made it my study and felt as if I had been to a university." A complete syrup and sugar maker comprises in himself a woodcutter, a forester, a botanist, an ecologist, a meteorologist, and [an] agronomist, a chemist, a cook, an economist, and a merchant. Sugaring is an art, an education, and a maintenance. (p. 246)

This paragraph by the Nearings overflows with the diversity that is quietly, silently imprisoned by our modern conception of the mundane as a knowledgeless land. The multiple roles we all play in our everyday actions, in contrast to the overspecialized view we most often have of our lives, speak most pointedly of the lost, colonized voices of know-how. How dare we consider ourselves botanists, economists, mechanics, or electricians! Yet, within our knowledge of the mundane, we often *act* and *do* as specialists, but we are not allowed to claim such knowledge because (in most cases) we were not taught such knowledge in a formal educational environment. Our own knowledge of know-how and use most often lies outside the bounds of school, in a hinterland ruled by the colonial hegemony of "those who know." We learn of know-how and use *through* practice, *so that the practice defines the theory* of our actions: the actions of know-how and use.

We learn as we do within the context of know-how and use; the actions beget the learning. The reversal of theory *then* practice to practice *then* theory is a table-turning phenomenon unrecognizable by many academic disciplines that instead champion the knowledge of theory over the knowledge of practice (if indeed the knowledge of practice is even recognized in the first place). Such a radical view of knowledge and learning, as that which is derived from practice, is an activity of reinventing not just the esoteric arguments of theory and practice but is a reinvention of the fundamental material makeup of our very educational systems. What would it mean to our educational institutions, the user might ask, if we made the knowledge of know-how visible within the confines of the academy? To paraphrase Aristotle, would the student judge better of the classroom than the teacher? Would the teacher judge better of the classroom (especially the computer classroom) than the architects who designed it?

Recovering the Mundane

As I have attempted to uncover these stories and tales for myself, as part of an effort to understand what it means to be a user of technology, I have from time to time used my own experience as a guide. Not unlike Thoreau or the Nearings, I have reflected upon moments that help me conceptualize what it means to use and to be a user. These personally situated experiences have enabled me to recall some of the lost voice of the user, and for a brief time I would like to indulge you in a few of the tales that follow:

I don't think that I could have been much more than ten or eleven years old, but the memory is nevertheless pungently clear. I was standing on the corner of Fifth and Broadway in Gary, Indiana (the town where I was 'born and bred' as they say), waiting for my father to come out of the building where he had an office. As I waited, I watched a man dressed in a doorman's uniform step from the front door of the First National Bank with a large push broom in his hand. Once out on the sidewalk, he began sweeping and continued to do so until he had whisked a significant amount of white-gray, dusty material out to the curb. He then pushed the dusty residue down the length of the sidewalk, off the curb, into the street, and finally into a storm-sewer grate where it fell quickly out of sight. The doorman returned to the main entrance of the bank, and with the broom still in his hand, held the door for a customer who stepped out onto the temporarily clean sidewalk.

Not long after the workman was done sweeping, my father appeared and we began walking to our car. On the way, I asked my father, "Why was that old man sweeping the sidewalk in front of the bank?" "He does it to keep people from tracking the dirt into the bank," my father replied. "It helps to keep the carpets in the bank from getting dirty so fast." Still not completely satisfied with the answer, I continued, "Why does the bank sidewalk get dirty so fast?" To answer this question, my father stopped, turned, and pointed his finger toward the north—directly at the main "Works" of U.S. Steel that lay a scant five blocks away. "You see the smoke coming from the "Works?" There's a lot of dust and dirt in that smoke, and it falls like rain on the downtown sidewalks every day and night. It's especially thick when water is dumped on the hot steel after it comes out of the blast furnaces. The man at the bank is kept pretty busy keeping that dust out of the bank lobby."

Just then, I saw a large white-gray cloud appear over the "Works," and it was followed by a muffled roar. "There . . . there it is now. They're pouring the water on the hot steel—thousands of gallons of it. There will be plenty more dust for him to sweep soon enough," he said as we turned back in the direction of the car. As we continued down the sidewalk, I noticed that the sky was changing color, to a sort of white-gray.

Several years later (actually, it was more like twenty) I recall standing in a century-old log barn with an elderly man who had once farmed the surrounding land. He was showing me some nearly antique farm implements that sat illuminated only by dusty rays of sunlight filtering through the boards of what had once been the hay maw. At one point he walked over to a horse-drawn grass cutter that was slightly sunken into the dirt floor. "This was the hardest implement for the horses to pull," he said. "It always took two of them to make it go. As you can see, it's all made of metal. Except, that is, for the wooden bar that sits in front of the blade."

I looked down and saw a wooden shaft, about four feet long, that connected two pieces of metal at the base of the cutter. "That shaft controls the sideways cutting action of the blade. They always had to make it out of wood. Hickory, usually. Sometimes ironwood. I remember how the companies that made the cutters would try to use different kinds of metal, but for some reason the metal shafts would bend, or worse yet, the other parts of the mower would get damaged. The wooden shaft, though, would just break...but even that didn't happen too often. It seems that the moisture content of the grass has something to do with it. The wooden shaft would "give" a little with the different moisture conditions and protect the other parts of the cutter from serious damage."

A few years later still (now I'm showing my age), I was teaching my first day of a writing class in a newly opened computer facility. Personal computers were still new, and the units in this room were even stranger because they were Apple Macintoshes®—all the other labs on campus to this point had either IBM-compatibles or mainframe workstations. So, we were all learning about how these icon-driven gadgets worked. Near the end of class, a student asked me to help her get the floppy disk out of the machine. "On the IBMs," she said "you just push that little button. But on these there isn't one. How do you get it out?" I reached for the mouse that sat next to her computer, and using it to move the little arrow on her computer's screen, I dragged her disk's icon to the trash can icon in the lower right-hand corner. "Oh, no!" she cried out. "Don't do that! I don't want to get rid of the disk, I just want to get the disk out of the machine." I assured her that I wasn't destroying or "trashing" her disk, but that the preferred method of ejecting disks on the Macintosh is to drag the disk icon to the trash. "As silly as it may seem," I said, "this is how you are supposed to do it."

Before she finished packing her backpack, I noticed that she put the disk back into the machine briefly to, I suppose, check that the disk was indeed unharmed. "Well," I thought to myself as she left the room, "we both learned something today about what it means to mix metaphors."

User knowledge is always situated. By that I mean what users know about technology and the experiences they have with it are always located in a certain time and place that changes from minute to minute, day to day, era to era. Hence, the complexity of understanding what users know grows with each new experience or story that we tell or hear. At the same time, however, there are connections and commonalities between these experiences that help thread them into a visible, knowable tapestry. In the first scene, the *user* is difficult to identify, at least as far as most common perceptions of users are concerned. Usually we think of users as being those at the end of the technology development cycle who take a tool, use it to make something, and then go on from there. In this scene, all three participants—the doorman, my father, and myself—are users of different types. Within the limited context of this brief scene, the doorman is, fundamentally, a user of two technologies—the push broom and the door of the bank. My father is a user of the office building and its accompanying technologies. I represent a user of a most ubiquitous technology—the sidewalk.

All three of us, however, are commonly connected through the use of technologies that we hardly notice, or if we do notice, then we do so with little or no reflection. When I spoke with my father, we were using possibly the most fundamental technology of the human experience—language. As unreflective as we may be of language as a technology, it is still a human construct, a human invention that is taught, learned, and used in strategic ways, much as we might use hammers, automobiles, or computers. In addition, all three subjects in this scene are users of the larger, more ominous industrial technology of steel manufacturing. Although we were not directly employed by the steel maker, we nevertheless used the products, the city, and even the economy that had been constructed by the technology of steel.[2] Operating unreflectively, we were (most often) unconscious users of this greater technology, and we literally or figuratively kept the sidewalks clean in homage to the technological "hand that fed us."

---

2. I mean most literally that in the situation of Gary, Indiana, the town and its people were very much "products" of U.S. Steel. The city was named for the superintendent of U.S. Steel in the beginning of the present century. The city of Gary did not even exist until 1906 when U.S. Steel decided to locate a mill there. The town was planned from the very beginning as a "steel town"—with the streets even mapped out years in advance of their construction in expectation of the "boom" that the area could expect. Today, like the American steel industry itself, Gary has declined in economic status to the point where the mills only employ a quarter of its peak employment of the 1960s, and the population of the city has actually declined in the last two decades.

In the old barn, I find myself more reflective of technology, possibly because it was placed into a historical context: a technology that "had been." In this scenario, technological use is imbued with a certain reverence lacking in the steel town image. That reverence, however, is not one of sentiment for lost arts or knowledge per se. I, of course, found the horse-drawn implements fascinating, in part because they represented a mystery of use that I could no longer capture in the present-day context. In other words, there were no longer any horses on this farm to move the implements, and the machinery took on more the quality of a snapshot or an artist's still life. Nevertheless, I have been continually fascinated by some technologies that defy theoretical answers. Some technologies are best understood only in the active context of their use. In the case of the horse-drawn grass cutter, the wooden shaft represented practice exercising a sort of victory over theory. No matter how hard the engineers tried to make a theoretically unbreakable shaft from the same material as the rest of the cutter (i.e. metal), they failed because the practice, the use of the machine, always demonstrated the best solution. I believe the old farmer held the same respect for the importance of practice, as he told the story of the wooden shaft with a certain glee in his voice.

Finally, in the computer operator of the third scene, we have a user and a situation of use that is easier to identify (and maybe identify *with*). Ironically, this young woman was using what has become the epitome of "user-friendly" technology—the Apple® Macintosh computer. Based on the familiar metaphor of the modern office, the Macintosh interface attempts to replicate an environment of use that anyone (who knows the essential material items of a modern-day office) can maneuver with ease. For the most part, the success of the Macintosh interface has been great. It does make computers more approachable (and it is easy to see why Microsoft Corporation wants to imitate that interface through its heavily marketed interface for IBM-compatible computers). But there are complexities of some technologies that even the most well conceived metaphors cannot overcome. Situated in the complexity of mundane experience, always constrained by the users' previous knowledge and the time pressures of everyday moments, technologies are constantly tested and refigured by those who use them. In essence, users understand technology from a unique perspective constructed from knowledge of practice within certain contexts. Yet, as de Certeau and a few others claim, this type of knowledge is subverted beneath a discourse of expertise, and thus has been rendered invisible to the modern eye. We take for granted that which we *do* and unwittingly surrender knowledge and power due to our lack of reflection on our mundane interactions with technology.

In so doing, we also surrender fundamental democratic rights and responsibilities. In a society that is perpetually bombarded with new technologies, it is important to reflect on what it means, in terms of the larger social order, to be a user of technology. I have found no better description of this very point than in the words of Aristotle that begin this chapter and set the tone for much of this book. The quote is taken from the *Politica,* and it is the conclusion of a discussion by Aristotle on the basic human right to take part in the decision-making activity of a culture. In a very practical sense, who has the right to voice an opinion or a vote? Who, Aristotle asks, is the better judge—the one who makes the product or the many who must use it? "Moreover, there are some arts whose products are not judged of solely, or best, by the artists themselves, namely those arts whose products are recognized by those who do not possess the art; for example, the knowledge of the house is not limited to the builder only. . . . " (*Politica,* 1282a 16–19).

The product can be a house, a ship, an automobile, a computer, or a vacuum cleaner, but in a larger sense the product is also the *culture* that we make through the arts and artifacts of technology. Users of a culture, in other words, often *are* the better judges, but if they are silent or invisible then they (we) have little power to affect the decision-making processes. We become passive observers of a social order that has the potential to be, ironically, undemocratic because it is governed by the technology that we have created. In *Critical Theory of Technology,* Andrew Feenberg describes the consequences of this phenomenon of an undemocratic technological order most succinctly.

> At the highest level, public life involves choices about what it means to be human. Today these choices are increasingly mediated by technical decisions. What human beings are and will become is decided in the shape of our tools no less than in the action of statesmen and political movements. The design of technology is thus an ontological decision fraught with political consequences. The exclusion of the vast majority from participation in this decision is the underlying cause of many of our problems. (p. 3)

Of course, the answer to this problem of citizen (user) participation and control is no clearer today than it was in Aristotle's time. In fact, it may have become more complex because the technological issues have become more widespread and powerful. But there is a common ground despite the separation of over twenty centuries, and it is situated, in part, in an understanding of how we, as users of technology, are empowered by virtue of our role as users. In other words, to revalue and refigure technological development and use from a user's perspective is a radical act, but one that has a strong historical base. The pages that follow are but a modest step in the direction of such a revaluing and refiguring.

Theorizing the Mundane

The narratives one collects or overhears concerning the everyday world are helpful in uncovering and situating the problem of mundane experience. The tales are interesting, revealing, and, I think, accurate lenses through which you can look at the complex variety that constitutes the ordinary world. But as much as the stories aid us, they fall short of providing a conceptual view of technological use that is fruitful in any systematic way—such as technical communicators might ask for. Consequently, what is called for is a theory: a theory based on user practice as it is associated with technological use. This theory of users and technology also must be cognizant of the social context—the cultural ambiance[3]—that ultimately situates the user and the technology. For the basis of this theoretical enterprise, we will now turn to the concept of user-centeredness.

What is meant by user-*centered*? The term that constitutes the title of this book has entered our present vocabulary through the context of computer technology, most visibly through the efforts of Donald Norman and Stephen Draper in a collaboration that began over a decade ago.[4] The exigency of their user-centered project was rooted in the multitude of problems people have when they use computers. In the introduction to their edited collection, *User Centered System Design,* Draper and Norman state that their " . . . book is about the design of computers, but from the user's point of view. . . . The emphasis is on people, rather than technology, although the powers and limits of contemporary machines are considered in order to know how to take that next step from today's limited machines toward more user-centered ones" (p. 2). They offer an invitation to those interested in working on the issues of user-centered system design: "We are prepared to take on board any discipline, any approach that helps. It is a pluralistic field. . . . " (p. 2).

---

3. "Cultural ambiance" is borrowed from the historian of technology John Staudenmaier. According to Staudenmaier, all artifacts are affected by the social sphere, the cultural ambiance, thus making technological artifacts and systems dependent upon, instead of autonomous of, human intervention. See *Technology's Storytellers.*

4. See their edited collection, *User Centered System Design: New Perspectives on Human–Computer Interaction,* listed in the bibliography.

In this book, I take the project of user-centered design set forth by Norman and Draper as a starting point, but I consider my project, although related, considerably different. For example, I emphasize that the problems associated with technological use are, literally, ancient. Historical context, in other words, is lacking in most user-centered research. The ancient Greeks, from whom I draw a number of concepts regarding technology and use, treated technology as an *art* whose *end* was in the *use* of the product, not in the design or making of the product itself. For example, in Aristotle's previously mentioned definition, the user is contextualized, recognized as residing in a situation of use where a special knowledge—the knowledge that users bring to interactions with technological artifacts—is championed. How often would we consider, in modern times, the user of a house more expert than the builder of the house?

Obviously, the user may know little, if anything, about the construction of the house, yet there is a knowledge—a knowledge of *use*—that is seldom acknowledged in most modern contexts. The specialist who designs or constructs the artifact is almost unquestionably the one who "knows" the artifact in our present day, and computer technology is no exception. The computer is presented as a black box technology that is driven by an expert/novice binary of knower/nonknower. To borrow from a post–World War II colloquialism, the computer in the black box view *is* rocket science in the sense that only a few experts can really understand how it functions. Computer users, thus, are more often viewed as idiots who must have the technologies "dumbed down" to their level, a level that has no knowledge of its own—only that knowledge that is handed down by those who made the object in question. It has only been recently, just the last few years, that those involved in user-centered design have begun to express interest in contextualized and situated interpretations of technology and use. We have much to gain by investigating ancient and indigenous conceptions of technology, most especially in user-centered quarters where such research has been virtually nonexistent. These connections and disjunctions between ancient and modern will be central to the discussions that follow.

Instead of focusing solely on computer technology as most user-centered design has done, I will expand the scope of the user-centered project to technology in general.[5] This I have done for several reasons. To begin, I first came to user-centered concepts through computer technology, specifically through user documentation. As I completed my dissertation, *Rhetoric*

5. User-centered concepts often have been used to provide analogies from contexts outside computer technology, but the actual applications of user-centered design in most other areas is work that is waiting to be done.

and Use: Toward a Theory of User-Centered Computer Documentation, however, I began to imagine what the implications of user-centered approaches to technology could more generally mean. It appeared to me that although computer design is certainly amenable to user-centered concepts, it was possible that some of the difficulties that users have with other types of technology were being overlooked. The glamour of the computer was possibly diverting the eyes of the beholders away from more plain, but nevertheless widespread, technologies of mundane experience. Thus, I was led to think about user-centered concepts as they might apply to a greater variety of technologies.

Second, the issue of context became more important as I contemplated what it means to be a user of technology. For instance, the complexity of multiple roles that one plays while using technology can be studied in a variety of environments, not just that of the computer. There are virtually an infinite number of domains that contextualize technological use, and the opportunity to investigate those realms was too enticing to ignore.

Third, user-centered research has dedicated itself almost entirely to probing the interactions that humans have with technological artifacts (usually computers), and therefore it is limited to the conception of technology as a material artifact. My interest in technology encompasses the discursive, or nonmaterial, aspects of technology and technological use. In particular, I will pursue the problem of *technique*—a term definitely related to the modern sense of technology but quite different when refigured through ancient concepts.

I also take Draper and Norman's invitation to "come on board" to heart, for there are no technical communicators present in their original collection. As I have already indicated in the Preface, conversations about technology and its cultural/political impact have been widespread. Philosophers, rhetoricians, linguists, computer interface designers, historians, sociologists, and literary scholars have all been part of this conversation (some, of course, more than others). Technical communicators have been on the periphery of these discussions, and our role in the dialogue has not been as visible as it should be. My project is aimed at entering this conversation and playing a part in furthering technical communication's engagement with the important social, ethical, and political arguments that are currently being discussed by other disciplines in relation to technology.

Technical communication already functions interdisciplinarily—this is one of its strongest assets—but interdisciplinary research is a two-edged sword. On the one hand, traversing various disciplines gives us a strong and penetrating perspective. It allows us to build theoretical, historical, and methodological bases in rich and invigorating ways. Like transfusions of essential nutrients, an interdisciplinary focus keeps us refreshed and alive.

It is not surprising, then, that technical communicators have borrowed extensively from a variety of disciplines in order to develop theories, research methods, and even the practical "tools of the trade" that are used in both the workplace and classroom. For instance, we have used theories of problem solving, social interaction, cognitive development, and visual design; we also have borrowed methodologies from ethnographers, linguistics, rhetoricians, and psychologists. We have grown immeasurably as a result of these borrowings.

On the other hand, interdisciplinary work has pitfalls which, although they often can be overcome, nevertheless need to be addressed. The list of these interdisciplinary pitfalls is potentially long, but one is most relevant to the context of this book: the problem of becoming so dependent on the borrower that we fail to reciprocate back into the interdisciplinary milieu with any contributions of our own . To be blunt, technical communication has not managed to give much in return, or maybe I should say that we have not been influential in having our voices heard across the terrain of the disciplinary divides. Maybe this is because of the "infancy" of our field; we have not established our identity strongly enough to be credible in the sense of academic disciplinarity. Technical communication may not even be established enough to call itself a discipline. Rhetoric and composition, for instance, was merely a service-oriented subfield of English studies until researchers established a theoretical and historical base for that emerging discipline. Technical communication may have a way to go before it can have the confidence brought about by disciplinary identity.

At the very least, our inability to reciprocate might be due to our ongoing interest in practical solutions to problems that technical communicators face on a daily basis: solutions that are not easily valued by some disciplines that consider themselves more theoretically or historically grounded. Engineers, for instance, perceived as mere practical artisans in the nineteenth century, sought the approval of the more respected theoretical "scientific" disciplines through borrowing scientific methods and developing measurable professional standards as avenues to disciplinary viability.[6] Whatever the reason for our reluctance to engage in these discussions, we should be more involved in a formative, interactive dialogue with these other disciplines—for us and for them. After all, we have plenty to offer.

In the chapters that follow, I present a modest offering to this interdisciplinary dialogue through an elaboration of user-centered technology in a

---

6. For more on the professionalization of engineers, see Edwin Layton's *The Revolt of the Engineers*.

variety of contexts. To ground this discussion, the next chapter will present a rhetorical framework of user-centered technology: a framework that imagines users as a viable part of technology design, development, dissemination, and use.

CHAPTER 2

Refiguring the End of Technology

Rhetoric and the Complex of Use

Every time you use a technical artifact, whether it is simple or complicated by nature, you use it within a "complex of use." For instance, if you are hammering wooden two-by-fours together to build the walls for a house on a commercial construction site, then your action of hammering is only one part of the larger complex. You, as the person doing the hammering, are acting as one of the creators, or artisans, of the artifact (the house). The hammer is the tool that immediately engages you with the artifact you are making and thus serves as a connecting point between you and the artifact, a kind of interface so to speak. You might be a new person to the trade of carpentry, or you might be an experienced carpenter who has built many houses and other structures made of wood. It also is possible that you are an intermittent helper who sometimes works as a carpenter, and at other times maybe you work as a forklift operator or a dump truck driver. Thus, as you actively build the house, you also are learning or doing, or both, as you go about your everyday business. As an employee of a company, you are part of a larger organization that may include managers, owners, unions, or other supervisory entities. Finally, you are situated in a certain cultural and historical moment that constrains everything you do.

The *end* of all of this activity, however, is not included in the aforementioned description. The *end*, from a user-centered perspective, is in the *use* of the house by those who will eventually live in it, namely the *users*.

In *Talking Power*, Robin Lakoff begins a discussion of the paradoxical nature of language by claiming, "Language is powerful; language is power. Language is a change-creating force and therefore to be feared and used, if at all, with great care, not unlike fire" (p. 13). Lakoff's statement is important in several ways to the central arguments of this book, as we will soon see. At the moment, however, I would like to concentrate on just one portion of her statement—the analogy between language and fire.

As you will recall, in Aeschylus' story *Prometheus Bound*, the titan Prometheus is condemned to eternal torture by Zeus because he had stolen fire and taken this elemental force of nature to the humans. It was not the fact that Prometheus had taken fire itself to humanity, however, that so angered Zeus: it was, instead, that Prometheus had shown humans how to harness nature and use it for their own ends. That is, Prometheus had brought to humans, through fire, the knowledge of *art*: the systematic, creative knowledge of craft and technique. Interestingly, Prometheus claims the art of language as one of these great gifts

> No token sure they had of winter's cold,
> No herald of the flowery spring or
> Season of ripening fruit, but labored without wit
> In all their works, till I revealed the obscure
> Risings and settings of the stars of heaven.
> Yea, and the art of number, arch-device,
> I founded, and the craft of written words,
> The world's recorder, mother of the Muse.

According to the Promethean myth, the use of fire was the first example of craft or technique to be experienced by humans. For Zeus, this gift giving was not perceived as being a friendly act as the application of fire to human ends was potentially the beginning of the end for the secret powers of the gods: the powers of technique (the ways of making) and technology (the artifacts that come from the making). The torrent of techniques and crafts that followed from this first example of fire promised to provide humans with Olympian powers of systematic, strategic knowledge. In short, Prometheus had given humans the power of knowledge, and one of the strongest forms of this "crafty" knowledge was language.

There was a price to be paid by humanity for this new knowledge, however. As Lakoff's statement points out, language is not just power but it is also *powerful*. Language, and by extension, technology, is a force that, like fire, can be used for burning, but it can also burn those who use it. Language and other technologies are tools that can serve practical or mundane functions like making, shaping, or fixing. But language also is power. It can persuade, control, and manipulate. When language is power, its usefulness is altered from a relatively neutral activity to an activity charged with opinion and controversy Thus, language becomes situated in what Western culture most commonly calls the realm of politics. Consequently, the *power* of language and other technologies is useful, but with that power comes responsibility for, and a respect of, the *powerfulness.*

Since the time of the ancient Greeks, this paradox of language has been the disciplinary domain of rhetoric; rhetoric has historically been interested in language as a tool *and* as a political force. As a result, rhetoricians have been compelled to probe the power and the powerfulness of language—a chore that has brought them into intimate contact with the social, ethical, and moral dimensions of language. In the following pages we will explore more fully the intimate connections between rhetoric, language, and technology. For now, however, what better place could we begin our investigations of "the power and the powerfulness" than within the domain of rhetoric?

Technology and Rhetoric: A Connection of *Ends*

Technology and rhetoric have a kinship whose genealogy can be traced back at least to the pre-Socratic Greeks, or to about 500 B.C.E. This long relationship has had a profound effect upon the development of Western thought and culture, and as you might expect, this long association has created some significant parallels between rhetoric and technology. Indeed, this is true, and throughout this book we will encounter a number of the commonalities between these two important forces of Western culture. Let us begin by focusing on one of these connections—namely their *end,* or what the ancient Greeks referred to as their *telos.*

Technology and "Interested" Ends

In the modern age, technology often has been seen as a panacea, a solution to many of the problems humans must solve and the hardships they must endure. Usually associated with the concept of progress or continual growth, this optimistic view of technology has piqued our imagination in countless ways. Americans "settled" the West with the aid of technologies such as the steamboat, the railroad, the rifle, and maybe the least complex, but nevertheless one of the most powerful, the steel plow; South and Central Americans have "tamed" much of the rain forest and the vast river areas of their continent with bulldozers, fences, pesticides and herbicides, and, of course, the railroad too; Europeans have "conquered" their forests with chain saws, controlled their rivers with dams and canals, and even populated barren stretches of Arctic tundra, thanks to technologies as different as power plants and snowmobiles. Many tales and stories, whether they are imaginative, historical, or journalistic, have romanced such accomplishments over the years and have done so in the name of "progress," or, in the name of the "good of humankind." In other words, the *end* of technology has been to move constantly, consistently, toward what we might blatantly and plainly call the "Good."

At the same time that we have romanced the progress and the "Good" of these accomplishments, we also have heard the cries of those who caution and worry about the "juggernaut" of technology. From the nineteenth-century paintings, poetry, and prose of Goya, Baudelaire, and Mary Shelley to the science fiction, ecological writings, and social critique of twentieth-century writers like Ray Bradbury, Rachel Carson, Wendell Berry, and Lewis Mumford, we have heard warnings of what technology is doing to our personal and social lives. In other words, the *end* of technology has been questioned extensively, and in particular we have asked, "Is the *end* of technology actually in the "Good," or is the *end* something more self-serving, more invidious, more 'evil'?"

This dichotomy of technology as being either "good" or "evil" is certainly reductive and simplistic, and the intent of the chapters to come will be, in part, to complicate this essentialist binary. Yet, there is a grain of truth in this fundamental schism, just as there tends to be in most paradoxes: We are enamored of the things that technology can promise, but we simultaneously live in fear of the power that unchecked growth and dissemination of technology has over our lives. We want technology to help us get where we want to go, but we feel uncomfortable if we are unable to control the direction and speed of the journey.

This desire to have a hand in the direction and speed of technological development is a normal human impulse. Just as we want to control the speed of our car as we round a curve or temper the heat under a pot of water, we also want to control the "larger" powers of technology, like atomic energy or urban sprawl. But how is this possible? How can individuals, or small communities[1] have an effect upon the growth of technology—those technologies that we see as getting out of control or moving beyond reasonable limits?

An answer, I will argue, is found in a refiguring of the *end* of technology: a fundamental rethinking of where technology is going and how humans can monitor its speed and direction. In modern times, the *end* of

---

1. I am using community here broadly. I mean as in a group who shares a common living area, like a town or city, but I also mean as in the Kuhnian sense of an intellectual or disciplinary community. I also would include, of course, virtual communities (see Doheny-Farina, 1996; Gurak, 1997).

technology has been too often in either 1) the interest of the developers who hope to gain from it, 2) the interest of the disseminators who likewise hope to reap the fruits of its success, or (and this is the most difficult one to see), 3) those who develop and then release a technology into the public sphere with little or no concern about its *end* whatsoever. It is this neglect of the Promethean dichotomy of the *ends* of power and powerfulness that I want to probe. Consequently, I propose that the end of technology be refigured as in the user: those humans (virtually all of us) who interact with various technologies (large corporate and governmental systems or small stand-alone devices; simple hand tools or complex electronic networks; discursively created or materially constructed artifacts) on a daily basis in our public and personal lives. To accomplish this refiguring, I will work in this chapter through the discipline of rhetoric, drawing upon its long history and rich theoretical base. Beyond this chapter, I will draw on a variety of other disciplines too—human factors, technical communication, composition studies, and the history, sociology, and philosophy of technology—to build the user-centered case. To begin, I would like to briefly discuss a few aspects of the history of rhetoric for the purpose of further demonstrating some parallels it has with the problem of technology's ends.

Rhetoric and the End of an "Art"

In modern everyday usage, the term *rhetoric* often has a negative connotation, as when someone refers to "the Senator's rhetoric," or when we label someone's speech as being "just rhetoric." This common, limited definition of rhetoric nevertheless has a long history. Usually traced to the works of the ancient Greeks, especially the dialogues of Plato, this connotation for rhetoric as empty words has laid a foundation throughout Western history of rhetoric as the art of deception: the instrument of devious and self-serving individuals who use language to achieve their own selfish ends. Socrates' descriptions of rhetoric as "cookery" in the *Gorgias*, or as the use of language to deceive in the *Phaedrus* are but two examples of this fundamental skepticism regarding rhetoric.[2]

Rhetoric, however, has another meaning—one both positive and powerful. In this uplifted sense, rhetoric is the *art* of creating (inventing), arranging, and delivering language for the purpose of evoking action upon

---

2. In the *Gorgias* I am referring to the moment when Socrates tells Callicles that rhetoric is like cookery in that it has no knowledge or epistemology of its own, but merely copies what is already known; the example from the *Phaedrus* refers to the example of selling a mule to someone who has been convinced through rhetorical means, that they have really bought a horse.

the part of an audience. That is, rhetoric is a systematic series[3] or collection of techniques that makes the production and dissemination of language strategic for the orator or writer, and, due to its systematic and thus transferable nature, is teachable to others.[4] Consequently, rhetoric has the power to persuade and distribute that power of language to others—something that made rhetoric highly compatible with the ancient Greek ideals of democracy and deliberative systems of justice.[5]

As you might surmise, however, a key difference between these two definitions of rhetoric (as deceit or as the strategic application of language) is, like technology, based upon the perceived *ends.* When the end is deceit or deception, the possibility that rhetoric might be used for unethical purposes presupposes that the rhetor will only use rhetoric toward his or her own gain. The definition of rhetoric as strategy, however, clearly defines the end of rhetoric much differently because, in this sense, rhetoric is defined as an *art.*

> The end of an art is *not* [sic] a product, but the use made of an artistic construct. The end of the art of housebuilding, for example, is neither the builder's use of the art nor the house itself, but rather the use made of the house by those for whom it was constructed. Similarly, the end of rhetoric is an active response in the auditor not the speech itself. (Lauer and Atwill 1995, *Refiguring Rhetoric*)

---

3. *Series* is meant in the sense of process, like the rhetorical canon where a rhetor moved through a process of inventing, arranging, styling, memorizing and delivering a speech. See the introduction of Bizzell and Herzberg's *The Rhetorical Tradition* for an extended explanation of the rhetorical canon as process. Herzberg's *The Rhetorical Tradition* for an extended explanation of the rhetorical canon as process.

4. A distinction between the classical and modern definitions of art might be helpful here. In the modern sense, art usually refers to the fine arts, or the artifacts of creative activity like paintings, sculptures, or poetry. The ancient term (and one used in some contexts today) refers to systematic techniques or crafts that an artisan uses to create an artifact. An example would be, "There is really an *art* to that!"

5. In Jean-Pierre Vernant's *The Origins of Greek Thought,* he explains that "The advent of the *polis* constitutes a decisive event in the history of Greek thought. . . . [The] system of the *polis* implied, first of all, the extraordinary preeminence of speech [rhetoric] over all other instruments of power" (p. 49).

Drawn from the Aristotelian concept of productive knowledge,[6] this definition of *art* places an ethical and a moral responsibility upon the rhetor/maker/artisan to make artifacts that suit the needs of the audience or, in the case of technology, the user. Hence, the end of rhetoric as *art* is in the hearer, or as the analogy to housebuilding demonstrates, the end of any kind of human activity involving making or producing artifacts (whether material or discursive) is in the receiver or user of the product. To use the analogy of the house, if the house would collapse upon the user of it, or if the user simply finds the house "unusable" due to poor planning on the part of the builder, then the builder (maker) has created an unethical product. The connections between rhetoric and technology in the context of *art* go considerably deeper than we have room for here, but nevertheless some further explanation is in order.[7]

In the Aristotelian definition of productive knowledge, the concept of art is referred to as *techne.* Techne, according to Aristotle, " . . . is identical with a state of capacity to make, involving a true course of reasoning. All art is concerned with coming into being, . . . art is concerned neither with things that are, or come into being, by necessity, nor things that do so in accordance with nature (since these things have their origin in themselves)" *NE* 1140a10–20. In the *Posterior Analytics,* Aristotle further explains the concept of techne as he compares it to one of the other types of knowledge in his taxonomy, theoretical knowledge [*episteme*].

> From experience again—that is from the universal come to rest in its entirety in the soul, the one along side the many, the unity that is a single identity within them all—originate art [*techne*] and science [*episteme*]: art in the realm of coming to be, science in the realm of being" (2.19.100a6–9).

---

6. In "Refiguring Rhetoric as an Art: Aristotle's Concept of *Techne,*" Janet Atwill and Janice Lauer describe the epistemological taxonomy of Aristotle as having three parts: theoretical (episteme), practical (praxis), and productive (techne) knowledge . They go on to explain that "the taxonomy explicitly locates the study of mathematics, the natural sciences and philosophy in the domain of theoretical knowledge, ethics and politics in the domain of practical knowledge, and the *arts* [our italics] in the domain of productive knowledge. Throughout the corpus, medicine and housebuilding are used as examples of arts" (p. 29).

7. Some of these connections will be discussed again in chapters 4 and 5.

Here, *techne*, the root of the modern term *technology*,[8] is associated with two issues vitally important to user-centered theory. First, *techne* is aligned with a "true course of reasoning." Thus, *techne* has the capability to create knowledge and allow for knowledge—issues of epistemology that will be addressed in depth in the third chapter concerning user knowledge. Such a connection between *techne* and technology will be helpful to build a case for the knowledge base of both technology and users.

Second, *techne* is not concerned with issues of certainty. The problem of certain knowledge, or *Truth* for Plato, is within the domain of scientific and theoretical knowledge (episteme). This is crucial because it is important that the definitions of knowledge we use to build user-centered theory are capable of dealing with contingency and mutability, or, in other terms, the "reality" of human contexts. Users, being human, operate in a world where things are constantly changing, constantly "coming into being."[9] Thus, technologies must be described or explained through a lens of contingency, probability, and/or mutability that accounts for shifting contexts and situations. *Productive knowledge*, particularly in terms of *techne* and rhetoric, provides such allowances.

We can see, even at a brief glance, that technology and rhetoric have had a close association, for good or bad, over the greater course of Western history. Prometheus' gift of fire was a blessing, but a blessing replete with warnings of the power that resides within tools—whether material or discursive. The problem of *ends*, in particular, has caused consistent debate from era to era inevitably affecting the power and status of the two respective concepts. Most pointedly, when *ends* are selfish—devoid of, or at least not centered on, human concerns—then the products that result often

8. While I wish to connect some senses of these two terms, *techne* and *technology*, I do not want to equate them synonymously. Such connections and disjunctions will be addressed throughout this book. However, let me say that I agree with such philosophers as Carl Mitcham (*Thinking*) and Langdon Winner (*Autonomous*) that these two terms can hold quite different meanings in their respective historical contexts, despite the fact that they also have some considerable connections.

9. Disenchantment with the idea of immutable truth is, of course, a central issue of postmodern philosophy. A refiguring of knowledge through productive knowledge is one way we can understand technology and language in a postmodern context. Such concern with the mutability of existence, however, is not just a twentieth-century invention. I am reminded of, for instance, Montaigne's essay "Of Repentance" where he discusses the mutability of experience and how it impinges on every aspect of life, and the final book of Spenser's *Faerie Queene*, the so-called "mutability cantos."

have an absence of essential communal human qualities, such as those of ethics and social responsibility.

In the rest of this chapter I will continue to discuss how the development and dissemination of modern technologies have suffered from this problem of ends. I will begin by demonstrating how modern technology is for the most part system- or artifact-centered. I will conclude with a new model of technological development—a user-centered model based on rhetoric, with users as the "end."

The System-Centered Model of Technology

The system-centered view is based upon models of technology that focus on the artifact or system as primary, and on the notion that the inventors or developers of the technology know best its design, dissemination, and intended use.[10] In general, the system-centered view holds that the technology, the humans, and the context within which they reside are perceived as constituting one system that operates in a rational manner toward the achievement of predetermined goals. As one researcher in the field of computer science puts it, system-centered approaches to computer technology have ". . . their historical basis in the idea that the ways of thinking used in computer programming can be used also when dealing with systems consisting of both human and other components" (Bødker 1991, p. 111). Another example of this approach to technology design is the cognitivist argument that the computer is a metaphor for the human mind (Boden 1988). This system-centered approach to artifacts and systems has dominated technological development at least since the advent of the Industrial Age, and only recently have significant critiques of this view been offered.[11]

There have been, of course, tremendous technological accomplishments over the last two centuries that attest to the apparent success of the system-centered view. The aforementioned examples of economic and

---

10. There is an interesting parallel here between system-centered views and romantic notions of authorship where the writer is seen as the primary agent, the creative genius of discourse. For ancient conceptions of the romantic concept of authorship and genius, see Cicero's *De Oratore*; a modern discussion can be found in Peter Elbow's "Eyes Closed."

11. We will investigate these critiques more pointedly in the upcoming chapter on technological determinism, but here I will mention that John Staudenmaier's *Technology's Storytellers*, Langdon Winner's *Autonomous Technology*, and Bijker, Hughes, and Pinch's *The Social Construction of Technological Systems* discuss this issue extensively. Donald Norman's *The Design of Everyday Things* and Hubert Dreyfus' *What Computers Can't Do* also offer interesting insights on this topic.

social expansion in the Americas and Europe illustrate this accomplishment. In our present time of electronic technologies, the increased speed of information processing and the development of tremendous memory storage capabilities are just two examples of computer technology's advance through system-centered design. Often, system-centered philosophy is determinist, as it perceives the system as an inevitable result of the logical progression of human activity or, in some even more extreme views, as the rational outcome of life itself so it is tantamount to natural law. One such position is presented by the economic historian James R. Beniger in his book *The Control Revolution*:

> . . . living systems have thus far evolved four levels of programmable structures and programs: DNA molecules encoded with genetic programming, the brain with cultural programming, organizations with formal processing and decision rules, and mechanical and electronic processors with algorithms. This succession of four levels of programming, each one which appears to have complemented and extended more than superseded already existing levels, constitutes the total history of control as we know it—a relatively smooth development punctuated by only these four major revolutions in control. (Beniger 1986, 61–62)

In the perspective Beniger presents, *system* is a metaphor for an essentialist view of life that moves from stage to stage on an ever-ascending ladder of progress and improvement.[12] To leash an understanding of the essence of all life to the metaphor of *system* is the goal of such a theory. It intends to model life as a naturally ordered, systematic phenomenon so we can, in turn, model our day-to-day activities in this "natural way." In representations of human life and our attendant technologies through the system-centered view, however, users are inevitably ancillary, or, in some cases, they are nonexistent because the system is powerfully hegemonic: the system is the source and ultimately the determiner of all. System-centered technology, as Figure 2.1 demonstrates, locates the technological system or artifact in a primary position. There is no need for the user to be involved with system or artifact development, this perspective suggests, because the system is too complex and therefore should be designed and developed by experts who know what is most appropriate in the system design. The system is created through a process of prototyping and iterative redesigning

12. Stephen Jay Gould offers a compelling critique of this "progress" view of history in *The Hidden Histories of Science, New York Review of Books*, 1995.

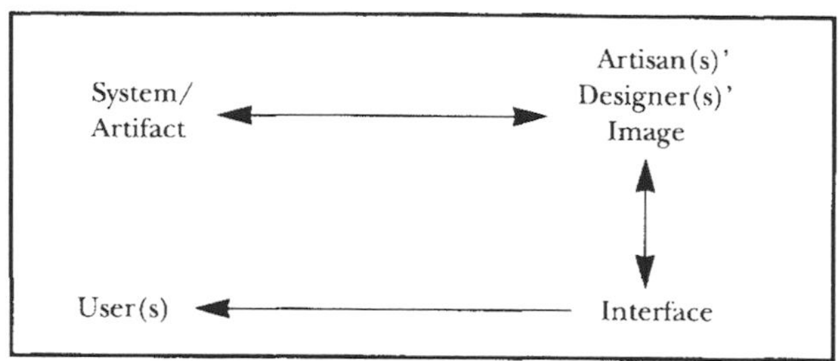

Fig. 2.1.
The System-Centered Model of Technology

that is primarily controlled by the designers or artisans. From this process emerges a technological artifact that embodies the *designer's image* of the system. Examples of this process have been told in many traditional historical accounts of the genius inventor working feverishly in a laboratory until the technology is "perfected," or the computer programmer writing code until the software is "debugged" and the functionality of the computer is complete.

From this designer's image follows the system *interface*—the "covering" of the internal system through which the user sees and operates. The interface, broadly conceived, can consist of actual "coverings," like computer screens or automobile dashboards, but they also can consist of the materials that explain a system's functions, like instructional materials or documentation. In what we often refer to as complex technologies, the interface can "cover" an enormous amount of information or machinery—such as in the electronic computer. The plentiful anecdotes associated with the difficulties of using some computer systems, like UNIX, are in large part stories of inadequate interfaces—images of the system that meant little to the uninitiated user. But even in less complex technologies interfaces can be problematic if they represent an image of the system that fails to meet a user's expectations. For instance, the common doorway can cause problems if the operation of the door is invisible or unclear to the user.[13] The interface is crucial to the user of the technology, but more often than not this intimate connecting point between the technology and the user is relegated to the end of the development cycle—at a point

---

13. Donald Norman provides an excellent example of this in his *Design of Everyday Things*. He shows how glass doors at a convention center are difficult to use because there are no clear markers to indicate the hinge side of the door, thus users walk into the unopened door. Similar issues are raised by Bruno Latour in his article, "Where Are the Missing Masses: The Sociology of a Few Mundane Artifacts."

where there is often little that can be done to solve any problems the user may have while operating the technology.[14]

In addition, interfaces can be merely complimentary or flattering to the user and therefore create an illusion of the greater workings of the system. These interfaces are commonly referred to as being *user-friendly*, and are not to be confused with user-centered designs.[15] User-friendly can describe a technological interface that is easy to use but may not necessarily be in the best interest of the user. In some cases and situations, user-friendly interfaces can be quite helpful. The Apple Macintosh® or Microsoft Windows® are two instances where user-friendly interfaces have helped make computers more usable for a greater number of people. At the same time, however, user-friendly interfaces like these can mask the complexities of the system to such an extent that if there is a system breakdown, such as when you receive a cryptic error message that explains the problem in virtually encrypted language (or you feel at least that it may as well be encrypted), you are left helpless, unable to solve the problem, and continue with your work because you are dependent on external expertise not available to you in any useful form.

In other cases, user-friendly technologies can be more problematic, even dangerous. The most mundane may be the electric light switch. You flip a switch and a light comes on. Simple enough. You have, however, through a simple, user-friendly interface just accessed a complex technological system that uses a vast array of natural, human, and economic resources in order to function. Every time we flip a simple switch then we are using a large, possibly controlling, technology; yet we are virtually unaware of the consequences in any immediate way. Even though we may read daily of the problems of overconsumption of electrical energy, we still are likely to (ab)use the technology because it is so "friendly"—so easy-to-use.

Finally, the user, far removed from the central concerns of the system or interface design, receives technology that is ultimately created in a system or system designer's image. Ironically, as the arrows indicate (see

---

14. Leaving users out of the development cycle has become one of the largest complaints of unusable systems (see Dohney–Farina 1992; Haselkorn 1988; Johnson 1994). Yet I have seen little evidence that technology developers are bringing users into these early stages of development.

15. Draper and Norman offer this definition of "user-friendly": "In practice it most often refers to verbose or chatty interfaces with long prompts, messages, and menus. While this may help some users, its benefits must be traded against the penalities of slow time displays, . . . distracting displays, . . . and degrading assumptions about the user" (p. 496).

Figure 2.1), it might even appear that the *end* of the technology is in the user because the user "gets it [the technology] at the end." Such a representation of the end is not what a user-centered theory would describe as the *desired* end. On the contrary, as far as the effect on the user is concerned, user-centered theory would agree more with a colloquial meaning of finally "getting it at the end" in terms of the system-centered model.

Consequently, due to the confusion or lack of acknowledgment concerning technology's ends, we continue to create technologies that baffle users and in the worst cases promote unethical uses of technology (Sedgwick 1993). Unfortunately, the system-centered view is so embedded in Western cultural ways of thinking about technology that even the best user-centered design approaches to technology can unwittingly fall victim to the system-centered ideology. Consider the following discussion of the user-centered work of Donald Norman.

Norman, in *The Design of Everyday Things*, is advocating a user-centered approach to technology design. During a discussion of user-centered design processes, he cautions that, "The designer must ensure that the system reveals the appropriate system image. Only then can the user acquire the proper user's model. . . . " While Norman is advocating that designers keep in mind the needs of the user, there is an interesting twist in his approach. He says that the user "must acquire the proper user's model" (190–91). This still implies that the system is at the center and that the job of the user is to learn what the designer provides.

An even more curious ramification is seen in the picture that accompanies Norman's description of how designers, users, and systems interact (see Figure 2.2).

Here the designer has the task of designing the system from a model of his or her creation. This model becomes the image of the system that is then passed on to the user. The user, as the arrows demonstrate, has the job of interacting with the system, thus he or she is expected to learn the system without any contact with the designer. There is, in short, no role for the designer after the initial development of the system. Why, it can be asked, does the designer not have feedback from the system, and more importantly, feedback from the user? It appears that the system is driving the user, and once again it serves as the central focus.

Norman continues that, "The system image includes instruction manuals and documentation. . . . Alas, even the best manuals cannot be counted on. . . . " (190–91). Unfortunately, the manuals Norman is undoubtedly referring to (although this is speculation because he does not cite them) are developed under a system-centered philosophy. Systems or the manuals developed from such a system-centered perspective "cannot be counted on," a truly user-centered approach would argue, because they are devel-

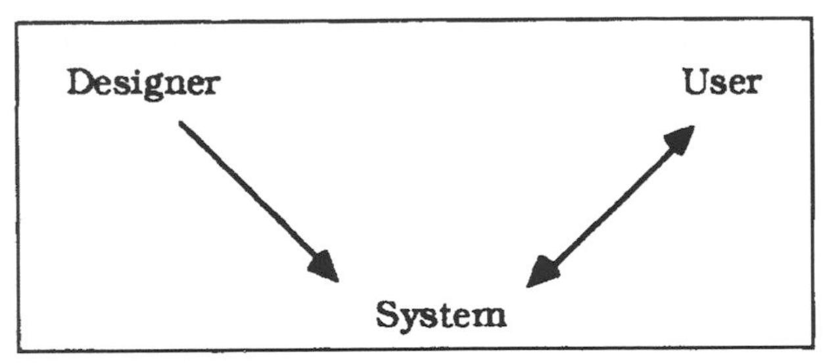

Fig. 2.2.
Norman's User-Centered Model

oped from a perspective that forces out an inappropriate image to the user. Users do not care about systems that reflect a designer's perspective: they want a system that is familiar and sensitive to their own perspective of the technology and its ends.

A theory of user-centered technology must keep the user's view of reality in mind to avoid entrapment in what could appear overtly to be user-centered ends, but could covertly actually be a reconstruction of system-centered ends. There is, in other words, no impact of the user's situation on the development of the system, and the result is that the system is often placed into the user's situation with the hope that it can be used.

The User-Centered View of Technology

The user-centered view is philosophically at the opposite end of the spectrum from the system-centered view. Principally, user-centered theory argues for the user as an integral, participatory force in the process (see Figure 2.3). By placing the user at the center of the model, the system and

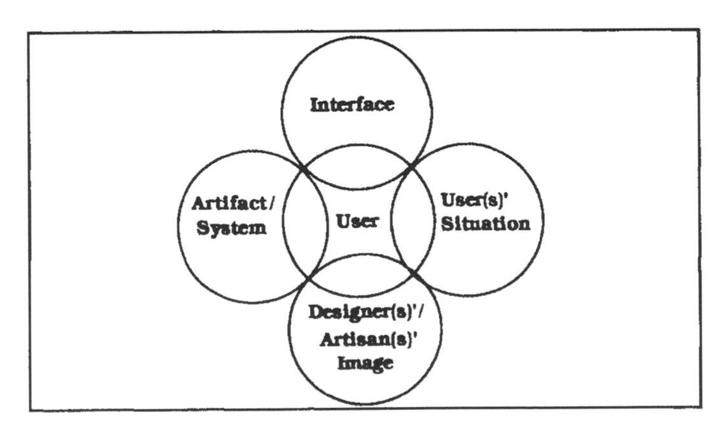

Fig. 2.3.
The User-Centered Model

the designer's image of the system are replaced as the dominant features. Instead, the model shifts the focus by concentrating on the user, and it adds another dimension—*the user's situation.*

The *user's situation* generally can be defined as the activity the user is engaged in and, in some cases, the moment in which this engagement occurs. Specifically, it represents the user activities of *learning, doing,* and *producing.* While these concepts of *learning, doing,* and *producing* will be explained in detail in chapters 3 and 6, suffice it to say at the moment that these are the general activities of a user's situation in that they define most of the global actions of users, at least as far as a user-centered theory is concerned. That is, when users are engaged in already implemented technologies (those technologies that are "complete" and "running") they usually are either *learning* how to use the technology, or they are *doing* something with it toward some end; when they are engaged in the processes of developing and maintaining technologies, they are *producing* technologies (they are, in other words, actually participating in the design, implementation, and maintenance of technology).

The user's situation also takes into account the *tasks* and *actions* he or she will be performing as a result of a particular situation of activity. This focus on tasks and actions within the user's context is clearly counter to the system-centered approach because here the use of technology is seen from the user's view of interaction with the technology, as opposed to a system-centered view that would describe these activities in terms of system features and functions.[16] Consequently, a user-centered view asks design and implementation questions like, What tasks will the user be performing within the given situation? How would the user represent these tasks within that situation? Are these tasks user tasks, or are they couched within the terminology or construct of system features? Are the tasks visible in the situation of use? Can the users, in other words, see what they are doing, or are the tasks and actions hidden behind an opaque or a clumsy interface?

For the most part, the descriptions of the remaining components of the user-centered model (the designer's image, the user, the artifact/system, and the interface) remain the same as they are in the system-centered model. A difference, however, (and this is a g*reat* difference) is how these constituent parts interact (see Figure 2.4).

---

16. An excellent discussion of this distinction between a user(s)' view of technology and the system view of technology is found in Dumas and Redish's *A Practical Approach to Usability Testing.* They explain that a system-centered approach describes the "functionality" of a system, while a user-centered approach describes the "usability" of the system.

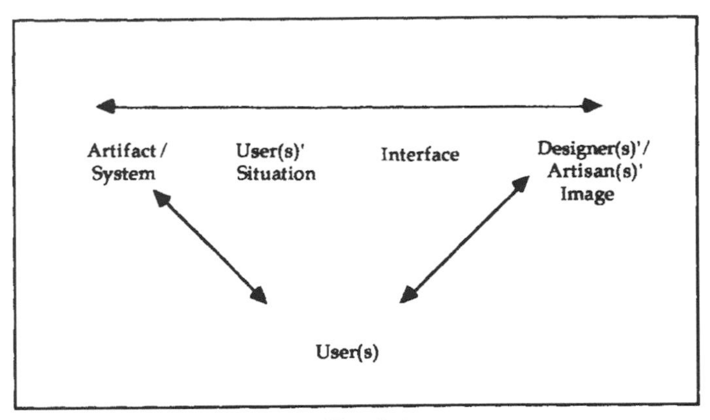

Fig. 2.4.
The Dynamics of the User-Centered Model

In a user-centered approach to technology, users are active participants in the design, development, implementation, and maintenance of the technology. This is not meant to imply that users are the sole or dominant forces in technology development. Rather, they are allowed to take part in a *negotiated process of technology design, development, and use* that has only rarely been practiced.[17] Users are *encouraged and invited* to "have a say," in other words, and thus they are physically or discursively[18] present in the decision-making processes of technological development. Invited to become actively involved with the technology in a greater, more integrated fashion, users become members of a team, but not just token members as they are in many usability approaches that merely invite users in at the end of the development cycle to validate a product. Because the users are involved with decision making in the user-centered model, they have power that historically has been concentrated solely in the province of the designers of technology (or in the province of the technological artifact itself, if you adhere to a strong determinist stance).[19]

---

17. There have been some significant successes that adhere more or less to this model. Possibly the most significant and successful of these are what has been referred to as the "Scandinavian Approach" (see Floyd, et. al.)—something we will see in chapter 4.

18. 'Discursively' means that the user may not be physically present, but that information collected directly from users (like interviews or surveys) is used directly in the development processes.

19. The issue of determinism and its role in user-centered theory will be addressed in depth in chapter 5.

The designer, as the arrows indicate in Figure 2.4, receives feedback about the technology from all quarters—the user, the interface, the artifact, and the user's situation. This embellishes the Norman model seen in Figure 2.2 in that the designer actually interacts with the user, not just the user as a mental construct of the designer or a component of the interface.[20] Also the interface of the technology is affected because it traverses all facets of the technology. Drawn from the system, the user's situation, the designer's image of the system, and from the users themselves, the interface—that crucial component of technology that users literally touch or feel—is derived from a true negotiation. Thus, collaboratively, the technology is created through a process of "give and take" that places users on a par with the developers and the system itself: a space within which users and developers can learn to value each other's knowledge and accept the responsibilities of technological design and development in new, shared ways.

Resituating the User: Rhetoric and the Complex of Use

Refiguring the place of the user within the space of technology can indeed be accomplished through rhetorical means, as we have just seen. The user, in this rhetoricized space, becomes an active participant who can negotiate technology in use and development. Yet, the schema of the user in this model is not complete because it is in large part unsituated. That is, the user-centered models just described in Figures 2.3 and 2.4 account for interaction among the various components of the technology design, development, implementation and end-use phases, but there also are situations that reside outside of these components that direct and redirect the interactions. The interaction among the various constituents, as it has been depicted up to now, is simplified in that the greater *complex of use* is not clearly exemplified. This final section of the chapter will complete the

20. The idea of the user as a fictional construct is an interesting one and one that has considerable merit. For example, Mark Simpson, in an ethnographic study of a computer publishing company, argues that the writers and editors "fictionalized" users through various role-playing activities (activities that were both conscious and unconscious for the writers). I agree that such fictionalized characterizations of users are possible and productive, but at the same time I contend that these techniques of user analysis are limited and are also driven more by economic realities (e.g., the expense of actually conducting user analyses or testing) than by actual concern for user input. I also believe that, as the Norman model in Figure 2.2 demonstrates, such fictionalized representations of users put too much power in the hands of developers (or the writers of the instructional documentation, as the case may be).

rhetoricizing of the user-centered model by 1) clearly defining the model as a rhetorical construct, and 2) situating the user-centered rhetoric within a complex of use: *the user-centered rhetorical complex of technology.*

Rhetoricizing the User-Centered Model

The concept of situation or context has been of ultimate importance to rhetoricians since the ancient Greeks, and its importance is no less recognized today. Ever since rhetoricians wanted to determine what was necessary to plan, arrange, and deliver a piece of discourse, they have developed taxonomies to aid them in their tasks. For instance, they categorized forums of discourse according to whether it was to be delivered in the senate, the courts, or in a ceremony. Similarly, they placed audiences into different groupings—young or old, hostile or friendly—to plan what would be appropriate for an intended audience, and then they assessed how they would persuade the chosen audience through a categorization of appeals (ethos, logos, and pathos). In the present century, there have been a variety of taxonomic devices that carry on this rhetorical tradition and one of the most well known is the rhetorical triangle. Early in the twentieth century, the philosopher and linguist I. A. Richards, while developing a theory of semiotics, used the metaphor of the triangle to describe what he termed the ". . . [e]ssential elements in the language situation [the symbol (words), the referent (things), and the thought or reference]," and the relations of these three elements ". . . [as] they are found in cases of reflective speech. . . ."(Bizzell and Herzberg 1990, p. 968). Thus, he formed a triangle to depict the relationship among these three elements, with each element residing on a particular point of the triangle.

In the 1960s, while working on the problem of describing the aims of discourse, James Kinneavy borrowed Richards' metaphor and revised it to explicitly fit the discipline of rhetoric. Kinneavy's triangle changed the terms on the three points from Richards' referent/symbol/thought to reality/reader/writer, and he provided a fourth term that was added to the center of the triangle—text (see Figure 2.5).

This metaphor works well as a tool for describing the components of the basic rhetorical interaction, but as one technical communication researcher has pointed out, the triangle does have a tendency to imply a static quality especially when it is used to depict communications as they actually occur in organizational settings (Spilka 1988). Additionally, there have been critiques of taxonomies such as the rhetorical triangle in general because they are reductive and thus can conveniently bypass the complexities of the communicative or ideological pressures that may impinge upon,

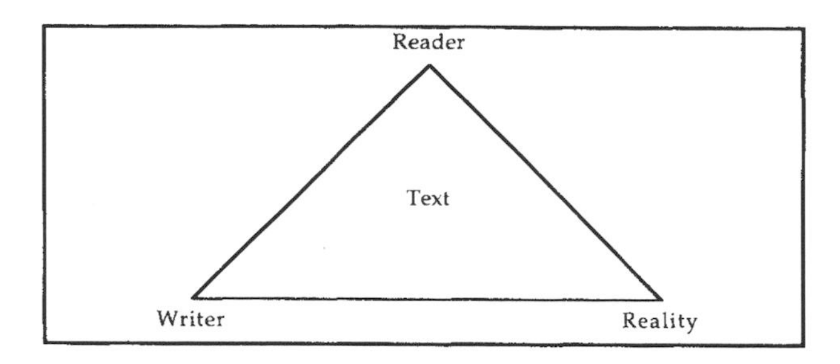

Fig. 2.5.
Kinneavy's Rhetorical Triangle

or are deeply embedded within their taxonomic representation. I agree that we should be aware of such critiques—they are correct in important ways.

In general, I believe we should be forthcoming about such inadequacies or dangers in any theoretical constructs we may make, and we should constantly reassess how the taxonomy may or may not be a fair or an accurate representation of our theory. Given an openness toward the inevitable reductiveness of the metaphor, we can move forward to discuss the theory in a forthright manner that reflects upon its deficiencies. While this is not a universal solution to the problem of taxonomic devices, it nevertheless makes one accountable for the ramifications of using such constructs.

In reference to the triangle as a taxonomic device itself, I think we can overcome the problem of such issues as the "staticness" of it as a visual aid by indicating the interchange that actually occurs.[21] In fact, Rachel Spilka does this as she adds arrows to the perimeter of the triangle to indicate the process of exchange that occurs. Also, there is an interesting irony at work as far as the reductiveness of the triangle is concerned. Few would argue that the triangle is not reductive: three or four points are about as simple as one can explain the communication process. Interestingly, though, the triangle has a certain "beauty" in its simplicity. By this I mean no single point on the triangle is more than one step removed from any other point. Put

21. It is important when using a construct like Kinneavy's triangle to keep in mind the original intent of the theorist. In Kinneavy's case, he was using the triangle, in part, to "break the lock" in composition studies regarding the modes of discourse. Most composition pedagogy from the nineteenth century is centered around writing instruction as a matter of learning modal techniques (narration, description, argumentation). Kinneavy used the triangle as a way to focus instead on the rhetorical aims of discourse (to persuade, to express, to refer). In other words, he wanted to move composition studies theory toward a model that concentrates on the needs of audience instead of the forms of text. See Kinneavy's *A Theory of Discourse.*

another way, unlike a square, pentagon, hexagon, and so on, each point on the triangle is always immediately in contact with the other two points: no point can change without having a direct effect on the others. Metaphorically speaking, then, the triangle represents an intimate connection between the various components in a way that would be virtually impossible with any other two-dimensional geometric construct.

The use of the rhetorical triangle in the user-centered theory takes considerable adjustment to the terminology itself and the placement of the terminology on the triangle. Figure 2.6 is a representation of the triangle to depict a user-centered rhetoric.

The writer of the Kinneavyean version is replaced by the *artisan(s)/designer(s)*. This, of course, puts the "creators" of technology into a terminology compatible with technological development. *Artisan*, for the most part, represents the maker of tools, or less complex (premodern?) technologies, while *designer* defines the more modern sense of the engineer or maybe even scientist (at least in a limited sense of scientist as a participant in the construction of technologies). At the same time that I have made these distinctions, I feel compelled to say that artisans often work with complex technologies, just as designers often work with simpler forms of artifacts. Later in this book, however, the distinction between these terms, either as a historical or cultural phenomenon, will be helpful.[22]

The text is changed to *artifact/system* and moves to the perimeter of the triangle. *Artifact* denotes technological constructs of lesser complexity[23]

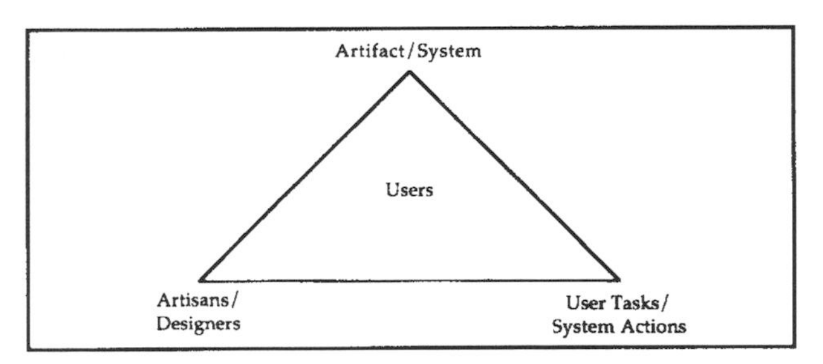

Fig. 2.6.
The User-Centered Rhetorical Triangle

22. Artisans and designers also include technical communicators, as technical communicators create technologies of language and other related products, such as computer interfaces. The technical communicator as artisan/designer will be broached in chapter 6, where issues of developing print and on-line computer documentation are discussed.

23. I hesitate to use the distinction "simple/complex" or "artifact/system" in some ways because I do not want to present artifacts (or artisans for that matter) as

and *system* denotes the more modern sense of technologies that intersect as a normal part of their "end." In addition, *system* will be used to discuss "nonartifactual" technologies, or what might be called discursive technologies, such as organizational structures or networks based upon different forms of communications.

What Kinneavy refers to as reality, or what some have called the "subject matter" of a communication act, is altered in the user-centered construct to the *user tasks/system actions.* This defines two ways that the "reality" of the technology is constructed and communicated. User tasks are the representations of the technology's actions as perceived by the user. System actions are the technology's actions as perceived by the artisan or designer. For instance, a computer programmer designing a drawing tool for making graphics might term a certain action as "specifying an element," but a user would see it as "drawing a circle," or "creating a shape." In other words, this key term in the user-centered rhetoric addresses the common breakdown people have as they move from the language of one community to another.[24]

The final component, the *user,* has replaced the reader in the traditional rhetorical schema. The user(s) is located at the center of the triangle as would be appropriate in a user-centered rhetoric. Like the artisan(s)/designer(s), the user(s) is plural to depict the collaborative nature of much technological use. Thus the user(s), in equal concert with the other elements of the rhetorical construct, completes the revised triangle. There is, though, one final stage necessary to complete the construction of the user-centered rhetoric, which entails complicating the rhetoric with the crucial elements of *situation* and *constraint.* No technology is developed, disseminated, or used in a vacuum, and a user-centered theory would be remiss, to say the least, if it bypassed this crucial concern.

---

being lower on a hierarchy. On the contrary, simpler technologies and the humans who use or make them are the most interesting to study in many contexts. My use of these two terms is probably best defined where *artifacts or simple technologies* are tools used independently of other tools, at least in any direct physical way, and *systems or complex technologies* are usually artifacts or tools physically connected either mechanically, electronically, or in some other direct manner. Alas, I think that coming up with new terminology for this distinction would be confusing.

24. The problem of distinguishing between designer and user language was previously footnoted in this chapter in regard to Dumas and Redish's "functionality/usability" distinction.

Situating the User-Centered Rhetoric of Technology

"Everything is relative," so goes the statement that has become a virtual truism in the modern world. As things move from one context or situation to another they are altered, reshaped, and refashioned by various pressures and constraints that exist in the new environment. Such "pressures," however, are not so much negative as they are formative: these constraints are what forge the user, the technological artifact or system, and the designers and artisans who take part in the enterprise. In brief, these pressures and constraints help form a *complex* that circumscribes, and, in turn, fashions, the rhetorical components of technology (see Figure 2.7).

In the first ring, the context of user activity is represented by the three activities of *learning, doing,* and *producing.* Described to some degree earlier in this chapter, these three terms define global activities that users are apt to be involved with during either the design, dissemination, or end use of technological systems or artifacts. Learning and doing are most closely associated with end-use activities and have been documented to a great degree by researchers in the many disciplines involving technology development and use.[25] Producing, however, is an activity rarely associated with users, and as discussed earlier, it is integral to a fuller understanding of the role of users in technological development. Users and producing will be described in more detail in chapter 3.

The next outward ring describes those constraints that larger human networks place upon technological use. These networks—depicted here as *disciplines, institutions,* and *communities*—probably do not constitute a complete list, but they nevertheless cover much of the ground. In addition, these networks could easily overlap and create complexes within/among themselves. One example would be the institution of the academy: various disciplines constitute its structure (e.g., the sciences, the arts, education, engineering, etc.), but there is also an overriding institutional structure that defines its purpose, objectives, proclivities, and so on. These activities all impinge on technology and ultimately on the various users of the technologies within the disciplinary and organizational boundaries. The term *community,* as well, is complex in itself, and here it can be used as broadly and completely as necessary. For instance, it can refer to well-defined communities like towns or organizations, but it also can refer to more ill-defined (or should I say less formal) constituencies, such as communities of discourse as discussed by social construction theorists.[26]

---

25. See, for example, the works of Thomas Sticht in the References section.

26. For a summary of communities of discourse, see James Porter's *Audience and Rhetoric.*

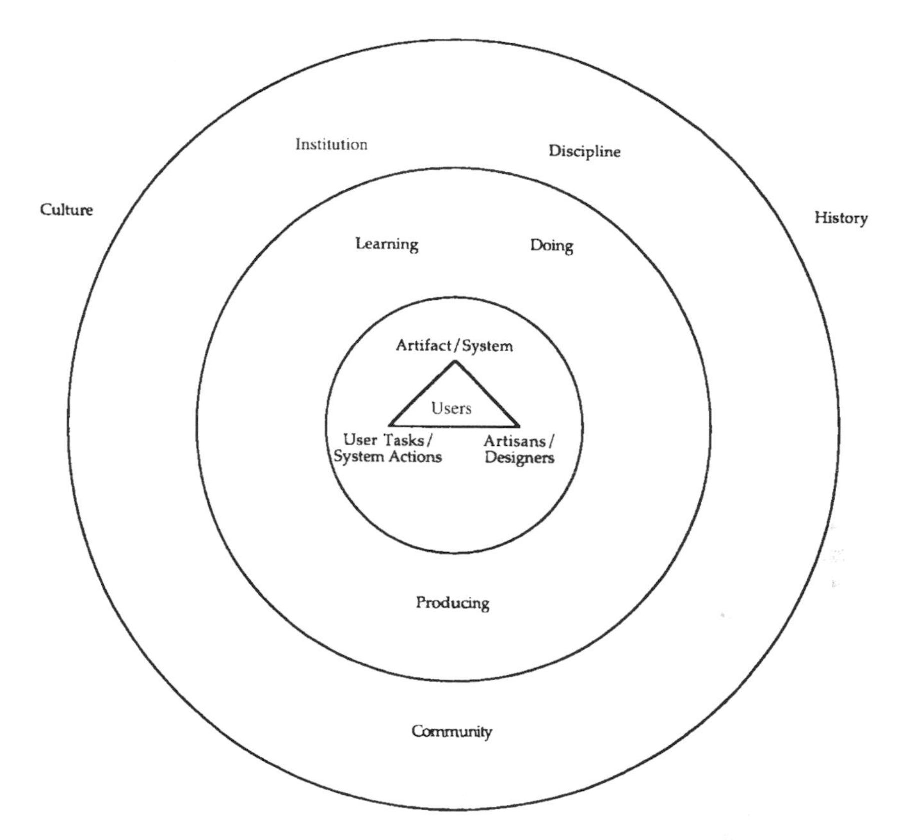

Fig. 2.7.
The User-Centered Rhetorical Complex of Technology

Residing on the outer edges of the complex are the factors of *culture* and *history*. These often invisible forces are nonetheless impossible to ignore (nor should any attempt be made to do so). Cultural forces define nearly every human action, and in a world more dependent than ever on international communication and technology transfer[27] the element of culture is without question essential when defining the use of a technology. History, although of course integrally related to culture, refers to the reflective aspect of understanding human action, particularly in terms of ethics and responsibility. Thus, history informs our understanding of technology in unique and indispensable ways.

---

27. For elaboration on these issues, see Stephen Doheny–Farina's *Rhetoric, Innovation, Technology: Case Studies of Technical Communicators in Technology Transfers.*

In conclusion, the complex serves a number of purposes. It can be a heuristic for analyzing technological artifacts or processes. It also can be a mode for exploring the people who use, make, and/or even destroy technology. It can help tell tales of people as they struggle with, become enamored of, or just get plain bored by technology. Finally, the user-centered complex of technology is a device for audience analysis that technical communicators can use, whether they are operating in the academic or nonacademic world, to study the audience we refer to as *users.* Consequently, we will put the *complex* to "use" in the following pages as we explore technology from a user-centered perspective.

In the following three chapters, we will continue to define crucial aspects of a user-centered theory of technology. In particular, we will ask questions that center on three of the essential points of the user-centered rhetorical complex—the artisan/designer, the users, and the artifact/system. In chapter 3, the problem of *user knowledge* will be addressed, thus we will concentrate on the user's role in the complex. Asking essentially, what constitutes users' ways of knowing, we will once again delve into the history and theory of rhetoric in search of a definition of this relatively unexplored territory of technological use. In chapter 4, the discipline of *human factors* will be historically overviewed to investigate the designer's perspective of user-centeredness. Human factors is where the term *user-centered* was coined, and thus it is appropriate that we investigate what that discipline means by *user-centered.* I will argue that although human factors appears to be user/human-centered, in reality it is more concerned with the designer's or the system's ends than it is with the user's. In the final chapter of this section we will concentrate on the artifact/system by confronting one of the largest and most difficult issues of the present day—that of *technological determinism.* Once again using inquiry as a method, we will pose questions about technological control and the possibilities that humans have for confronting the issues of (and irony of) "controlling technology."

PART II

Complicating Technology

CHAPTER 3

Not Just for Idiots Anymore

Practice, Production, and Users' Ways of Knowing

> Epistemological orders are, in a very literal sense, social orders.
>
> —*Janet Atwill, Instituting the Art of Rhetoric*

> . . . Almost all human actions fall within the province of chance (tuche). . . . But the fact is that although one may appear to be quite right in saying this about seafaring and the arts of the pilot, the physician and the general, yet there is something else that we must equally well say about these same things . . . namely that God and, with the help of God, Tuche [chance] and Kairos [the propitious moment] govern all human affairs; and that these two collaborators with God must be followed by a third which is our own [of humans], Techne. One must agree that to possess the art of steering ships rather not to possess it is our means of salvation when there is a storm. . . .
>
> —The Athenian stranger speaking in Plato's *Laws*

Several years ago, I gave a presentation at the Conference on College Composition and Communication where I spoke about the rhetorical nature of visual design (Johnson 1989). A main point of the talk was that users of technology have little time or patience for what we commonly term *reading*. Put briefly, my argument[1] was based upon the premise that users of technology are actively engaged with a technology that is taking all of their concentration, thus making another mental activity such as reading a near impossibility. I stated that users of technology, especially those learning a new technology, need to have instructional texts that are visually sensitive to the users' needs as they negotiate the activity of simultaneously learning the technology while "reading" the instructional text. During the discussion session following the talk, a gentleman sitting near the front remarked, "What you said about reading bothers me. I mean, if we just make the information easier to read, aren't we just 'dumbing it down?' Aren't we just making it too easy for people—so that they really don't have to be challenged to learn?"

My point in providing this anecdote is not to discuss the merits of visually designed text. Instead, for a moment I want to focus on what the statement about "dumbing down" implies about how we perceive users. Literally, it represents users as being dumb—as having no knowledge of their own. Therefore, this attitude implies, users must have everything provided to them in an easy-to-use format that asks little of them in return. In addition, the idea that users are provided with already digested knowledge relegates them to a subservient role: a role of mere practitioner, or someone positioned at the bottom of the proverbial epistemological ladder.

The idea that users are "mindless" is nothing new. The ancient Greeks placed those who carried out the activities of everyday existence (those who practice) below those whose daily life consisted of contemplation or observation (those who theorize).[2] During the Renaissance, we see more evidence of the mindless practitioner represented in craftspeople, like the "mechanicals" of Shakespeare's comedies who see life in one very literal dimension. In our own time, we need only think of the terminology of "idiot proofing" that has pervaded the design of technological systems to see the deeply entrenched notions we have of users. From automobiles to computers, the concept of idiot proofing has defined the view of the user:

---

1. This, of course, is not just my argument but is part of a large conversation among technical communicators regarding visual meaning. I cannot list everyone here who has done work on this topic, but a good introduction would be to look at Barton and Barton, Sullivan, Kostelnick, Horton, and Bernhardt. An excellent piece on the problems of reading and use is Patricia Wright's "Toward a Theory of NOT Reading."

2. See John Burnet, *Greek Philosophy: Thales to Plato*. He describes three stations based upon epistemology that the Greeks adhered to: the Theoretic, the Practical, and the Apostaulic.

someone who knows little or nothing of the technological system and who is seen as the source of error or breakdown. Our cars have "idiot lights" that tell us nothing of the underlying problem, and our computers have cryptic error messages that ultimately must be interpreted by the "knowing" expert. Users reside on the weak side of the idiot/genius binary. We have embedded the notion of technological idiocy so strongly in our culture that we actually begin to think of ourselves as idiots when we encounter technology breakdowns. Experts are the ones who "know," so we let them have the power, which of course means we accept whatever is given to us. As a consequence, when we watch our car being towed to the garage, or as we sit helplessly on the telephone help line, we often feel stupid. Users are not innocent in this enterprise of idiocy, though. Users themselves have in many ways allowed the construction of the idea of technological idiocy through an acquiescence to the knowing expert and to the acceptance of the idea that technology is just too complex for the "average" person to understand. Users are more than end-of-the-line consumers—they have a responsibility for the design and implementation of technology, too, as we shall see later in this chapter.

To raise users from the status of "idiots" will take some serious attention, and such attention will have to be more than just a raising of user consciousness by users themselves or by the designers of technologies. To effectively implement changes to current notions of users and use will take concerted efforts to understand the *knowledge* that users have of technological artifacts and systems. Even though there has been effort expended toward understanding essential characteristics of the users of technological systems, and in particular electronic technologies,[3] there continues to be a lack of serious attention paid to the nature of user knowledge. As Carl Mitcham explains in an excellent treatment of use and using, "Despite discussions of practice by Marx, Dewey, Polyani, and others, the philosophical analysis of using is slighted. Although the philosophy of action and ethics have something to say about using, the concept of use is conspicuous by its absence in all major texts" (*Thinking Through Technology*, p. 230). The reason for the absence of discussions of users and use runs deep in the history of Western culture, and at the root is the question of ownership of knowledge—in short, the question of epistemology. Who creates knowledge? Is it created only by those who we generally equate with knowledge, like philosophers or scientists? Or is knowledge production also within the

3. Here I am referring to the tremendous interest in usability that has been generated over the past two decades since the advent of the personal computer. See Nielsen (1992) and Dumas and Redish (1994).

province of those generally associated with "the practical," such as the technicians or users of technology?

The implications of these questions are crucial to a user-centered theory because how they are answered affects not only the validity of the theory, but more important, the answers will define crucial aspects of the social and political status of users. Such a questioning of epistemology can affect a refiguring of how we view the cultural world. We will be calling into question basic assumptions of hierarchy, power, and control: assumptions so entrenched in our culture that we rarely, if ever, question their presence. Put another way, refiguring technology onto a user landscape forces a rethinking and potential revaluing of material, social, and political relations in radical ways.

In this chapter we will start to answer some of the questions pertaining to user knowledge by mapping out characteristics of *a user way of knowing*. In particular, I want to focus on three aspects of user knowledge. The first of these aspects is *user as practitioner*. This view takes for granted that users are merely using an already implemented technology toward some fairly well-defined end. In other words, users as practitioners usually are portrayed as the users of tools: users of ready-to-go artifacts that have already been designed and handed to them for some particular purpose. We also will see how there is a paradox concerning users being mere practitoners. On the one hand, when users are viewed as only the mere implementors of technology, there is little room for a user epistemology other than as an "idiot" who receives technology and then puts it to use. On the other hand, there is knowledge involved with practice that should be highly valued. In particular, there is a *cunning intelligence* involved with practice that has been virtually overlooked.

The second aspect of user knowledge to inspect, and one that has been virtually nonexistent in discussions of use, is the *user as producer*. Here the user not only "practices the technology" but also is involved with the knowledge that constructs the technological artifact or system. That is, users as producers are capable of being designers and maintainers of technology: humans who are important factors in technological decision making (as opposed to the unfortunate human factors we will see exemplified in traditional human factors research in chapter 4).

The third area of user knowledge we will consider is the *user as citizen*. We will investigate how users, particularly in a democracy, can serve as active participants in the larger technological order. This final aspect of user knowledge will invest users in the social fabric of technology as being contributing, and equally responsible, members of the technological enterprises of our culture.

Knowledge of Practice: User as Practitioner

Possibly the most common representation of a user is that of a tool user. As such, the user is seen employing a technological artifact to create or make another artifact. The examples one can generate of tool use are conceivably countless, but they can range from the use of a fundamental hand tool (like a hammer, broom, or pencil) to a more elaborate, albeit probably hand-driven, tool (like a computer, radar control panel, or telephone switchboard). This representation of users and their knowledge of technology is helpful for elaborating user knowledge because it presents a predominant impression we have of technological use.

A tool use model of user knowledge provides a sense of one way that users know technology, and indeed an important way. At the same time, it can be shortsighted in that the model relegates user knowledge to mere how-to.[4] Almost as though some greater, more knowledgeable force had handed down the "truth" of technology to users, the assumption of the user as mere tool user virtually strips users of any epistemology whatsoever.[5] The user as tool user becomes no more than a mere practitioner. Further, in the tool use view, the user is seldom allowed a role in the production or maintenance of the technology. The tool-use model ultimately has the effect of describing user knowledge from a tool-centered, artifact-centered, or system-centered perspective, because the knowledge of the technology is assumed to be in the technology, not in the user.[6] Consequently, if we limit our definition of user knowledge to tool use models, we are acknowledging that the knowledge of the technology is put there by designers or inventors, not by users. A result of this one-dimensional view of user knowledge is the subjugation of users to the role of rote learners or awestruck spectators, thus perpetuating the view that users are idiots.

Langdon Winner's discussion of technology as a tool use model is helpful in further elaborating on this issue. In a critique of the tool use

---

4. Such work as represented by Adler and Winograd in their collection *Usability: Turning Technology into Tools* is tremendously helpful research that goes a long way toward bringing users into the design and development process. Unfortunately, I'm afraid that its title tends to further the assumptions I critique here.

5. The issue of technological determinism that is implied here will be discussed in depth in chapter 5.

6. It is curious that most philosophical conversation regarding users focuses on *use*, but is rarely, if ever, concerned with *users*. Thus, the discussions tend to usually center on the artifact as the thing being used, but rarely center on the user as being anything more than an appendage of the technical artifact. See pp. 230–240 of Mitcham's lucid discussion of technological use in *Thinking Through Technology*.

model, Winner claims that it is primarily a result of " . . . nostalgia. There was a golden age when the hand was on the handle and alchemy was queen of the sciences. But except for the world of small-scale appliances, that time has passed" (p. 202). He goes on to explain that the complexity and internetworking of technologies since the advent of the Industrial Age has changed not only how we view technology but how we interact and engage with it.

> We do not *use* technologies so much as we *live* them. One begins to think differently of tools when one notices that tools include persons as functioning parts.[7] Highly developed, complex technologies are tools without handles or, at least, with handles of extremely remote access. Yet we continue to talk as if telephone and electric systems were analogous in their employment to a simple hand drill, as if an army were similar to an egg beater. (*Autonomous Technology*, p. 202)

Indeed, technologies are becoming more complex, at least in the sense that many of the constituent parts of the artifact are either physically removed from the user (like computer networks) or are masked in such a way that the user sees only the surface of the technology (like computer interfaces or automobile dashboards). In the context of user knowledge, if these modern technologies are interpreted as tools, then the user has little hope of being involved with the greater knowledge of the technological system. Removed from the decision-making processes and design stages of how the tool will be constructed and what purposes it might serve, the user as tool user is a knowledgeless puppet whose only claim to epistemic status is the prescriptive knowledge he or she has of the use of the tool.

But is not this just exactly what the user of technology is supposed to be?: a user of already constructed artifacts (tools) whose expertise resides in the ability to handle a hammer, a computer, or an automobile with aplomb? Maybe. But in terms of knowledge creation, practitioners are rarely held in high regard. They are the borrowers of knowledge, a perpetuation of essential binaries that constitute the hierarchy of Western culture—theory/practice, philosophy/rhetoric, science/technology. Practitioners, in other terms, are nearly always situated on the weak side—the knowledgeless side—of a binary. Thus, to refigure users into an epistemic space of their own is to dispell the binary and, in turn, provide

---

7. One is reminded here of Donna Haraway's notion of cyborgs. See *Simians, Cyborgs, and Women: The Reinvention of Nature.*

users with knowledge creating powers that transcend the notions of mere practice.[8]

Let us begin by examining a scenario that 1) illustrates some key characteristics of this unique form of knowledge and 2) exposes some of the complexity that is commonly thought of as merely the practical, mundane, or everyday.

Imagine that you are at a self-service filling station and you have just finished filling your car with gasoline. It is the middle of winter and the temperature has dipped well below freezing, and in fact has been below that point for the last several weeks. You pay the cashier and hop into the driver's seat, only to find that the car won't start . . . not even a "whirr" from the starter or a "click" from the solenoid. It is completely dead. Over the next half hour or so a series of attendants and other customers inspect your car. They check the battery, bang on the starter with a wrench, they try to boost the battery with jumper cables. They even run a diagnostic on the electrical system with an electronic analyzer. Alas, it is all to no avail; the car will just not make a sound.

Just as you are about ready to give up and possibly pull out your credit card to buy a new battery (even though no one is convinced a battery will solve the problem, but what else can you do?), onto the drive comes an old car driven by a somewhat grizzled man wearing a mechanic's uniform. He gets out of his car, snuffs a cigarette under an oil-soaked boot, and walks over to your car. "Won't start?" he asks, knowing perfectly well the rhetorical nature of his question. You nod affirmatively, after which he proceeds to reopen your car's hood to survey the problem. "Has anyone taken the battery out?" he asks.

"No," you reply, "at least not here at the station." "So," he responds, "it has been taken out lately, right?" Now, feeling somewhat interrogated, you answer that indeed you have been taking the battery out every night because of the extremely cold weather. You explain that you've brought the battery inside at night and have been putting it back in every morning because the room temperature battery starts the car much better.

"A good idea, all right, to take the battery in at night when it's this cold. I often do it myself" he says, reassuring you that you haven't been "screwing up." Out of his coveralls he pulls a small wrench and proceeds to undo and remove both battery terminal ends. He then wipes off both the ter-

---

8. The theory/practice binary and its epistemological manifestations are discussed extensively in Pierre Bourdieu's *An Outline of a Theory of Practice*. See especially the introduction to his text and pp. 168–179.

minal ends and the battery posts with a clean, dry rag. He quickly puts the battery cable ends back on the battery posts, tightens the nuts, and then says, "Try it!" You turn the key and the car starts without any effort whatsoever.

Amazed, you ask him what he did. It appeared that all he did was wipe off the terminals. Was this some form of magic? "Like I said," he explains, "it's a good idea to take the battery inside at night. Only problem is that the warm battery can cause water to condense on the terminal ends. You essentially have created a barrier that the current can't pass through. So, the system shorts out. Just remember to wipe off the terminals before you put the cable ends on and tighten the nuts."

There are at least two types of users as practitioners in this scenario. One, the driver of the car, represents the vast majority of us who use technologies on a daily basis but who know little of how the systems function. The other user, the mechanic, is more knowledgeable about the technological system and in particular how that system (in this case, the car) operates within different situations or environments. Yet the mechanic is still a user—he merely interacts with the system in a different manner than the driver, usually as a maintainer of the system.

A difference between the two users could be characterized as "expert/novice" in terms of the technological knowledge they possess.[9] To a certain extent, this is an accurate distinction. The driver is something of a novice as far as the internal system is concerned (yet the driver does know how to remove a battery). However, the driver is probably an expert in terms of operating the system (e.g., driving the car), and the driver appears to have some knowledge of how to maintain a car in a cold climate. In contrast, the mechanic is an expert of the system, but he is probably an expert in operation of the system too. However, it might be possible that in another context (say if the mechanic moved to central Texas and took a job repairing cars in a hot, dry climate) the expert mechanic might become a novice (or at least a less accomplished expert) in some respects as he learns to negotiate the impact of a different environment on the internal components of automobiles.

---

9. The expert/novice distinction is a common one that has been used extensively in the study of human knowledge, particularly in reference to problem solving. See for instance the works of Herbert Simon, John Hayes, or Linda Flower, listed in the References section at the end of the book.

Clearly, the expert/novice distinction helps characterize some differences between these two users, but it is a problematic distinction as well[10] because it potentially creates hierarchies of knowledge, thus running the risk of marginalizing novices as users who must learn from, or be merely the observers of, the expert. The knowledge of novice users is more complicated than the expert/novice binary suggests. Novices do not necessarily need to become, nor do they often want to become, experts—at least in the sense of becoming a system expert. In many contexts, novice users are more like border crossers: individuals or communities that move in and out of technological contexts for different purposes and different ends.[11] Thus novice/expert distinctions only help tell part of the story, and we should find ways to explain these differences in richer terms.

A second aspect of user knowledge as practice that surfaces in the scenario, and one that begins to flesh out a richer sense of user knowledge, is the ancient Greek concept, *techne*—the knowledge of arts and crafts associated with the making of things. Techne is a complex concept—one that has been debated by philosophers and rhetoricians for centuries and subsequently has various definitions and connotations associated with it. One of the chief debates concerning techne has had to do with its own epistemic nature. In short, the question that has arisen over techne is, Is techne epistemic? Is there, in other words, a "knowing," or as the Greeks would have said, "logos" associated with techne?

The Greek term *techne* often is translated as "skill," "art," or "craft," and in the ancient texts it is associated with such activities as carpentry, navigating, or weaving. In one sense, techne is described as "knack" or "pure technique."[12] In this sense, techne is a systematic way of making, but the knowledge of the art or skill is tacit or unconscious. Thus, techne as "knack" is a definition of the human activity of making where the knowledge of a skill is the only requirement. In other terms, "knack" characterizes techne as a natural phenomenon—it is either something you have or you do not have and can only be handed down, most literally, as rote information. For a user, this definition of techne would be associated with the unreflective adherence to prescribed procedures. This pejorative, or at least devalued

10. For an extended critique of the novice/expert distinction, see Dreyfus and Dreyfus (1986). They characterize the distinction not as a leap from novice to expert, but rather as a five-stage process of development.

11. See John Seely Brown and Paul Duguid, "Borderline Issues: Social and Material Aspects of Design," and Lucy Suchman, "Working Relations of Technology Production and Use."

12. See Richard Young's "Arts, Crafts, Gifts, and Knacks: Some Disharmonies in the New Rhetoric."

sense of techne, obviously is related to the idea of the user as being a "mere practitioner" as mentioned earlier.

As I have discussed in chapter 2, techne is defined as an " . . . art [that] is identical with a state of capacity to make, involving a true course of reasoning. All art is concerned with coming into being, that is, with contriving and considering how something may come into being which is capable of either being or not being, and whose origin is in the maker and not in the thing made. . . . " (Aristotle, *Nicomachean Ethics,* 1140a10–15). This vision of techne is crucial to user knowledge because it firmly situates techne and any of the arts or crafts associated with techne, in an epistemic domain. The contention that the origin of an artifact is "in the maker and not in the thing made" shifts the focus of epistemology away from the artifact or system. Like the navigator of the ship mentioned by the Athenian stranger at the opening of this chapter, the user of the technology has, through techne, the knowledge to overcome the potentially inevitable consequences brought forth by domineering and determining forces. For the ancient Greek, these controlling forces may have been the gods that represented the overpowering forces of nature (like Tuche or Kairos). For us, these forces may be technology itself.[13] At any rate, techne represents the human force, the human knowledge that permits control *through* technology—whether the technology is a basic hand tool or an intertwined network of information services. Techne, as an element of user knowledge, goes a long way toward providing us with a more thorough understanding of, and basis for, user knowledge. Specifically, it offers a definition of user knowledge as a generative source of knowledge that gives users the potential for a significant degree of involvement and power regarding technology.

In the aforementioned scenario of the service station, techne is present in different forms. For one, the driver of the car uses techne (although probably unconsciously) through the action of taking the battery in at night. In this sense, techne is a learned strategy, a rule of thumb for solving a problem, even if the solution is only a temporary fix. This example of techne is useful because it exemplifies how a learned process of solving a

---

13. Some scholars argue that the ancient term *techne* and the modern term *technology* are linked etymologically, but the terms have quite different connotations in their respective historical contexts. In brief, ancient techne is associated with rhetorical invention, and thus is epistemic in that it helps a rhetor "come to know" what he or she must say. The modern concept of technology, however, is not necessarily associated with invention or knowledge and instead represents artifacts and mass production techniques. See Mitcham, *Thinking,* pp. 114–136; Winner, *Autonomous,* pp. 8–12 and 73–77; Dunne, *Rough Ground.*

problem is the result of a rational action: a "logos" of action. Users build a repertoire of strategies and, when needed, they draw upon this repertoire of strategies to solve a problem in a given situation.[14]

The mechanic provides a more detailed example of techne as he moves through a series of questions to pinpoint the problem at hand. Much of his techne is invisible to the observer, however, as his knowledge of the strategies he employs is tacit.[15] Although he makes the procedure of figuring out the problem appear simple, in reality his ability to solve the problem is in part the result of quickly eliminating many possible solutions. Guided by an analysis of the constraints of his present situation, the mechanic systematically "guesses" at the potential solution.[16] It is pertinent to point out here, too, that the procedures the mechanic used are the very things technical communicators often must document as they write the prescriptive "how-to" information of instructional materials. As we traverse the geography of techne later in this book, we will even be challenging assumptions of prescriptively designed user documentation as the best method to write procedural materials. In other words, refiguring the knowledge of technology through users' eyes challenges some of the most basic assumptions of technical communicators' writing practices, too.

In this scenario we also get a glimpse of what is probably the most unexplored, yet possibly the most powerful, aspect of user knowledge—the concept of *metis*. Metis, or what is also called cunning intelligence, is the ability to act quickly, effectively, and prudently within ever-changing contexts. Related to and sometimes described as a component of techne, metis is derived from ancient Greek mythology, as the word itself was the name of the wife of the first wife of Zeus.[17] A brief detour into the Greek myth of

---

14. This concept of techne and problem solving is quite similar to the technique of problem solving advocated by G. Polya in *How to Solve It*, and the concept of "satisficing" as described by Herbert A. Simon in *The Sciences of the Artificial*.

15. See Polyani's *Personal Knowledge*.

16. "Guesses" is in quotation marks here because I want to point out that the "arts of guessing" are important components of the intelligence of users associated with the Greek concept of *metis* (which will be defined shortly). Drawing from the Greek term *eikazein*, these skills of guessing are a stochastic intelligence based upon a "good eye" and the ability to see in front and behind at the same time. For more on this, see Detienne and Vernant, especially pp. 314–315.

17. I do not mean to imply here that the term or the concept of *metis* is "owned" by Greek culture. To be sure, the term *metis* is found in Native American legends and refers to the "trickster" or border crosser (sometimes the mixed blood human)—see Powell 1994. The connections between the uses of the term *metis* from culture to culture are unclear and would be an interesting and a valuable avenue of research.

Zeus and Metis would be helpful here in describing this unique quality of user knowledge.

In *Cunning Intelligence in Greek Culture and Society*, Marcel Detienne and Jean-Pierre Vernant tell the following story of the goddess Metis:

> The goddess Metis who might be considered a somewhat quaint figure seems, at first sight, to be restricted to no more than a walk on part. She is Zeus' first wife and almost as soon as she conceives Athena she is swallowed by her husband. . . . Not content to unite himself to Metis by his first marriage, Zeus made himself pure *metis* by swallowing her. . . . Henceforth . . . there can be no *metis* possible without Zeus or directed toward him. Not a single cunning trick can be plotted in the universe without first passing through his mind.[18] (pp. 11–14)

Literally, Zeus wanted to control *metis.* But why? Zeus—the most powerful god—had control over the heavens, but that control was tenuous as other gods constantly competed for the throne (a not uncommon problem for any king, whether mortal or otherwise). Yet it might be queried, why would Zeus prize so highly a form of intelligence associated with practice? Certainly, as we have already seen in reference to Plato, the Greeks positioned the philosopher (the contemplative theoretician) at the top of the epistemological ladder. It would seem only logical that theoretical knowledge, not practical knowledge, would be the essence of power.

Apparently, knowledge was valued differently in the ancient context even though a definite hierarchy existed then too. As Detienne and Vernant explain, "From Homer to Oppian practical and cunning intelligence, in all its forms, is a permanent feature of the Greek world. Its domain is a veritable empire and the man of prudence, of *metis,* can assume ten different identities at once . . . ranging from the charioteer to the politician and including the fisherman, the blacksmith, the orator, the weaver, the pilot, the hunter, the sophist, the carpenter and the strategist. He turns up everywhere. . . . "(pp. 307–08). Practical knowledge, especially knowledge of making aimed at some end, was seen as being very important to the ancient Greek mind. We seem to have lost this same value for knowledge of action and making. Even though some Greek philosophers, such as Plato,

---

18. The swallowing of Metis by Zeus also is a prime example of the issue of "invisibility." In particular, it is a literal statement of the dominance of a supreme or universal "knower" (Zeus) over a "knower" of contingent or local knowledge (Metis). In addition, feminist issues of invisibility are raised here quite explicitly. Like Evelyn Fox Keller's study of the geneticist Barbara McClintock, Metis (like McClintock) is rendered invisible by the dominant male power of Zeus (male scientists).

adhered to definite hierarchies of knowledge, they also valued the knowledge of practice and human action differently than we do now. In the modern context, we simply value theoretical or scientific knowledge more highly than we do technical or practical knowledge.

I, of course, am not alone in holding the view that theoretical and scientific knowledges are hegemonic. Many historians of technology point to the dominance of scientific method since the Enlightenment to explain this shift in the value of knowledge. The knowledge of science, in other words, creates the knowledge that is then handed to technology for application. Although this simplistic view of technology as applied science has been refuted consistently over the past forty years, it is apparent that the belief that scientific knowledge resides over technical or practical knowledge remains strong.[19] Interestingly, Detienne and Vernant see this lost value of practical knowledge as a driving question in their research on *metis*. In the conclusion of their book, they muse that, "It may well seem paradoxical that a type of intelligence as fundamental and as well represented in a society such as that of ancient Greece should have remained so neglected. It is all the more surprising in that the fourth century philosophers Plato and Aristotle certainly referred to it, describing its characteristics and defining its qualities" (p. 308).

They answer this inquiry, however, with two interesting and telling conclusions. First, they suggest that cunning intelligence was often equated with animal intelligence by the Greeks (e.g., the cunning of the fox; the continually "open eye" of the octopus). Thus, they speculate that the Christian ethic that places human over animal would not have found a form of intelligence associated with animals fitting for its own philosophy. Consequently, cunning intelligence was subverted by the controlling sweep

---

19. In the science versus technology debate, technology is relegated to a position of applied science while science controls the direction and course of knowledge. Most clearly articulated by the philosopher Mario Bunge, technology is ". . . the field of knowledge concerned with designing artifacts and planning their realization, operation, adjustment, maintenance and monitoring in the light of scientific knowledge" (1985, p. 231). Technology as applied science, then, has no real epistemic quality of its own as it can only borrow knowledge from, or test the knowledge of, science. In fact, technology as applied science leaves technology with no knowledge of its own except the practical know-how usually associated with keeping already constructed technologies "up and running." As far as user knowledge is concerned, technology as applied science strips all types of users (e.g., from drivers of cars to mechanics) of an epistemological base. For more on this debate, see Layton ("Mirror Image Twins"); Mitcham (*Thinking*, especially pp. 199–204); Kranzberg ("Unity").

of Christian theology in the centuries following the Greek and Roman eras. They follow this first speculation with a second and more compelling one for our purposes:

> The second and even more powerful reason is surely that the concept of Platonic Truth, which has overshadowed a whole area of intelligence with its own kind of understanding, has never really ceased to haunt Western metaphysical thought. (p. 318)

The predominance of universal truth and certain knowledge, they argue, has subverted the situated and contingent knowledge of the practical arts, like *techne* and *metis.* Subsequently, in Western culture we have constructed a philosophy of knowledge that not only devalues the practices of the everyday, but also devalues the knowledge of those who function in that context.

Back to earth and in the modern world once again, let us analyze *metis* operating in the filling station scenario. Metis is most visible in the mechanic as he quickly and deftly solves the problem of starting the car. The mechanic obliquely approaches the situation (he sizes up the situation and guesses at its nature), briefly asks questions that help dissect the problem, and then corrects the problem based upon past knowledge and a cunning analysis of the situation. Thus, the mechanic displays a "logos" or reasoned approach to the solution of the problem, yet he does so heuristically. The mechanic is flexible in his systematic approach to the problem and does not follow a set algorithm to reach a solution. He also displays prudence as he asks the question of the driver, thus wisely assessing the situation and at the same time including the knowledge of the driver in the decision-making process.[20] Although a small component of this particular scenario, it is nevertheless a visible example of how shrewd ethical awareness is an important characteristic of user knowledge because it demonstrates the value that can be placed on practical knowledge between different types of users.

Finally, through the aforementioned scenario, I want to emphasize the complexity of user situations. As we saw in the rhetorical complex of tech-

---

20. Prudence is used here in the sense of what the ancient Greeks called "phronesis" or practical wisdom. One who exhibits practical wisdom acts not just toward the goal of creating an end product but also with the intention of assuring that the action(s) are aimed at the "good." For instance, in the *Nicomachaen Ethics,* Aristotle explains that "Practical wisdom, then, must be regarded as a reasoned and true capacity to act with regard to the common good" (1140b17–18). Phronesis and prudence will be discussed in more detail later in this chapter under "User as Participatory Citizen."

nology (Figure 2.7), when users confront technologies there is a *complex* of events that occur (even though the events may appear mundane). Many technical communication approaches to understanding situations of use have been to look at one dimension of user interactions—usually the step-by-step procedures users carry out. Thus these situations are not perceived as complexes, but rather as individuals learning about technology in linear and simplified manners. The filling station scenario, to take just one instance, is a complicated set of social, technological, and knowledge interactions that are difficult to decipher as reduced moments of mere interaction between a user and a technological artifact. As you will remember from the user-centered complex of technology from chapter 2 (Figure 2.7), the user situation is circumscribed by a variety of constraints that play a large role in the shaping of any particular moment of technological use. In short, understanding user knowledge is not just a matter of simplifying the complex. Such reductiveness tends to depict the experience of users as being merely a mundane experience and can relegate user knowledge to the "epistemological basement," as we shall see both literally and figuratively in the following section, user as producer.

Knowledge as Production: User as Producer

Users, as the very term implies, are manipulators: literally, humans who manually operate some form of technological artifact in the practice of something like a trade, craft, profession, or (in the modern context) recreational activity. Through this definition, users are seldom if ever portrayed as being *producers* of knowledge. Instead, they are most often depicted as practitioners who rarely create new knowledge, but who may very well be "receptacles" of considerable amounts of information that has been provided for them—similar to the role of technology in the applied science model.

The view of users as being receptacles of information is highly limited and is also inaccurate. Users are producers of knowledge, but their modes of production have been rendered invisible by those modern cultural proclivities that subordinate the user to being a mere practitioner. The ancient Greeks, however, saw the practice *and* production characteristics of users as being important and embodied them quite visibly in their mythology through the goddess Athena—the daughter of Metis.

> Building and driving are two types of activities in which we are more prone to notice the differences than the resemblances. However, the ancient Greeks saw many affinities between them as is shown by a number of pieces of evidence concerning Athena. . . . The truth is that, where the horse-drawn chariot is concerned,

> Athena's activity is more complex than we first imagined: it is not restricted to driving the chariot and horses but extends to the building of the carriage and the putting together of the various pieces of wood from which the chariot as a whole is composed. . . . Whether a chariot, a plough, or a ship be involved, Athena presides over all the phases of working with wood—the felling of the wood, the planing of the planks and the fitting together of the different pieces forming the framework, for all of these are operations which demand an equal measure of *metis*. As Homer says, it is "not strength but *metis* that makes the good woodcutter.' And every carpenter is first a woodcutter: his initial task is to chop down the tree trunks that he himself has selected in the forest. (Detienne and Vernant, *Cunning*, pp. 234–235)

It is difficult for us to imagine in the present day how integrally related the arts of practice and production were for the Greeks. Once again involving the cunning intelligence of *metis*, the end use of technology for the ancients involved forms of knowledge underrepresented in our modern context. Users, in the ancient definition, know the "how" *and* the "that" of technology as it moves from context to context through iterative processes of production and practice. Yet the idea of users as practitioners and producers is not as far removed from the present time as we might suspect.

The previous quote reminds me of a box of old photographs I once found in the attic of a farm in the Upper Peninsula of Michigan. The pictures, taken around 1910, were of the original owners of the farm as they were building the house, barn, and outbuildings. Some of the pictures portrayed horses dragging white pine logs for the walls of the house and barn, while others showed members of the family hewing the logs, pegging them together with ironwood as they formed the walls and splitting cedar shakes for the roof. One of the photos showed an outbuilding that still stands and now contains a wide array of woodworking tools: crossbuck saws for felling trees; hammers, saws, and augers for building houses and barns; finer tools for furniture making. The house where I found the photographs clearly reflected the tools that remained in the outbuilding. Inside the knee walls of the second floor you could see the hand-hewn logs of the original exterior walls—white pine logs so finely hewed by hand with heavy broad axes that you could run your hand across their surface without fear of getting a splinter. The house, the tools, the craft were intertwined, meshed. The mix of practice and production—of artisan, user, and artifact—was essentially seamless in these photos that were less than a century old, and in the tools and artifacts remaining from that not-so-bygone time.

To be sure, the strong focus on specialization in the twentieth century has created a culture where we seldom think of making things from scratch.

Nowadays it would be more appropriate to describe forestry specialists planting and maintaining the trees, logging companies harvesting and transporting the crop, mills preparing the lumber into standardized dimensions, marketing specialists helping lumber yard managers to determine the best-selling products for their particular locale, and, finally, carpenters buying the lumber materials and building an artifact. It is easy to get nostalgic and sentimental about the loss of know-how, but such nostalgia usually becomes a reactionary call for a return to the past. Although I think it important to mourn the loss of generalist skills associated with producing something from scratch, it is more important that we *actively* pursue changes to the social order that carefully assess the realities of the present situation. The hand-wringing associated with nostalgic sentimentality solves next to nothing, as we have seen in recent calls for a return to basics in education, which merely advocate a return to a time that exists only in the minds of those who have benefited from the educational status quo.

Instead, we should bemoan the loss of a sense of values related to users as they are involved in the actions of practice and production. Consider the following scenario in a modern law office as an example of the ramifications of this loss of broad-based epistemological values:

A team of workplace ethnographers is nearing the completion of a lengthy study of a large law firm. The ethographers have discovered that the firm (not surprisingly) has a strict hierarchical structure that stretches from senior partners at the top to litigation workers at the bottom. During their investigations of the firm, the ethnographers have become well acquainted with the senior partners and the litigation workers. One of the ethnographic teams most interesting observations was how important the litigation workers were to the success of the firm. These workers, many female and without advanced degrees for the most part (although most had bachelor's degrees), spend their days researching the cases the lawyers are working on, compiling that research into meaningful structures, and ultimately preparing written products for the lawyers to use in their cases. In short, the ethographers came to realize that much of the knowledge the firm relied upon was created by the "low level" litigation writers.

Ironically, the litigation writers were in many ways "invisible" to the higher-level workers (even literally invisible, as the litigation writers were housed in the basement of the firm's building). The ethnographers, through interviews with and observations of the firm's lawyers, found that even though the lawyers constantly used the products of the litigation workers in preparing their cases, they never thought of the written products of the workers being in any way an act of knowledge creation. They merely thought of the litigation workers as scribes who put together already known information (drawn from a network of electronic databases and archives) that the lawyers wended into finely crafted arguments for use in legal proceedings.

Near the end of the project, the director of the ethnographic team was telephoned by one of the senior partners to discuss "a sensitive matter." In the conversation, the partner told the director that there might be some extensive layoffs among the litigation staff (in essence they were going to eliminate the whole litigation support staff), and he did not want the ethographers to mention that the layoffs were possibly going to happen. Surprised that the firm would be letting go of such a valuable asset, the ethnographer asked why they were doing this. The senior partner replied that the writer's services were too expensive, and with computer networks being what they are today, the firm can have the work of the litigation writers done more cheaply in the Philippines. The written products, then, would just be E-mailed back to the firm's headquarters in California.[21]

There are a number of tacks that can be taken regarding this scenario. Questions of ethics, downsizing, and the use of electronic networks are but a few issues raised by the case. For our purposes here, though, the issue of knowledge is clear: those who reside at the lower, more practical levels of social orders are the least appreciated. And the knowledge they do produce is not valued as being anything but mindless regurgitation of what is already known. In essence, the litigation workers in this scenario are like the proverbial cogs in the machinery of a modern technological system—they are central to the workings of the machine, but are nevertheless invisible and replaceable.

Imagine the law firm as a computer system where the lawyers are the "brains"—the mysterious but controlling knowledge base of the system. The litigation workers, who reside outside of the internal system, are the users of the computer system who operate through an opaque interface,

21. This scenario is taken from a description of a project conducted by ethnographers working for Xerox Palo Alto Research Center (PARC) in an article by Lucy Suchman, "Working Relations of Technology Production and Work." A telling description of the undervalued knowledge of the litigation workers is given by Suchman: "So we found ourselves [the ethnographers] cast into the middle of a contest over professional identities and practices within firms, framed by the attorneys as a distinction between 'knowledge work' [the work of the lawyers] on the one hand and 'mindless labor' [the work of the litigation workers] on the other, framed very differently by the workers within litigation support themselves" (p. 32).

essentially channeling information into a black box, into the brain. On the inside of the black box—the lawyers' side—decisions are made, functions are carried out, and the "real work" of the system is accomplished. Even though the litigation staff has specific knowledge of how the system operates and are important members of the knowledge production process, they nevertheless are invisible to the knowledge base of the system. In turn, the litigation workers ultimately become expendable because the work they do, and the knowledge they produce, is unseen because it passes into the black box and becomes part of the controlling system—a kind of colonizing effect. Consequently, the litigation staff becomes an invisible piece of the system. They are so subsumed by the technological systems—the organizational system of the firm *and* the computer technology they use to write and transmit their information—that the partners of the law firm see virtually no difference between the living, breathing litigation workers and the litigation texts that would be electronically transmitted over several thousand miles of ocean. Here, in an ironic turn on the postmodern conflation of writer/text, artisan/artifact, the already disempowered worker is further disempowered as he or she becomes an invisible part of the "machinery."

Waitresses, mechanics, secretaries, carpenters, bank clerks, even technical writers—all have valuable knowledge of the systems of which they are a part. Yet, they often are the most expendable, the least valued members of the knowledge system. Like the "idiots" who use technologies, those who hold practical positions in the hierarchy have the least power even if they are, like the litigation workers, actually producing knowledge that turns the literal or metaphorical gears of technology. We are obliged to learn how to value, how to see, the knowledge that users produce, and then, in turn, make this knowledge an integral part of the technologies we use, or are a part of each day.

Knowledge of the Polis: User as Participatory Citizen

Users as practitioners and producers are interesting definitions of user knowledge in themselves. Users can be systematic, cunning, and downright intelligent. These definitions, however, still remain somewhat unsituated; they are too abstract to truly allow user knowledge to have a powerful effect. Additionally, these two definitions, no matter how useful, do not ask enough of users. Users defined as practitioners and producers tells us much about what user knowledge is, but there is not much indication regarding how users can become active, responsible members of the technological community. For this, we must add a third component of user knowledge—*user as participatory citizen.*

In the rhetorical complex (see Figure 2.7), the user(s)' situation is circumscribed by institutional, and ultimately cultural forces that help shape the environment of the user. These same forces, of course, share in shaping the epistemology of users and the status they may or may not enjoy, as the case may be. The spaces in which the knowledge of users and the cultural environment intersect are difficult to describe, though, because there is no terminology for discussing users in large, social contexts. Whereas we have found ways to depict other epistemological realms where practice and production occur in a social context—artisans, designers, scientists, technologists, philosophers, rhetoricians—users, for the most part, remain just users. This is especially problematic because, as was already mentioned in the discussion of users as practitioners, users are most often depicted as being mere tool users—rarely a definition that invites an appreciation of social or cultural context.

To define the user as a member of a larger social context I have, as you might suspect, turned to the ancient Greeks. From them I borrow the concept of "citizen," or member of the *polis*. I will begin by discussing the ancient idea of the polis and then move to a discussion of users as citizens through a scenario from our modern day.

Scholars have described the development of the polis as one of the crucial turning points in western culture. Vernant describes the polis as a new "spiritual universe" that displaced the closed systems of information and power that had been under the control of religious leaders and aristocratic families who ruled the populace. This shift from monarchical and religious rule to a democratic order was, of course, not recognizable at any one moment. There were, however, several elements that we can decipher through our "historical gaze."

Vernant provides us with an explanation of three important characteristics that allowed the polis to develop. First was the "extraordinary preeminence of speech over all other instruments of power" (p. 49). Rhetoric put power into a citizenship to take part in decisions of the state. "Speech was no longer the ritual word, the precise formula, but open debate, discussion, argument. . . . All questions of general concern that the sovereign had to settle . . . were now submitted to the art of oratory and had to be resolved at the conclusion of a debate" (p. 50). Thus rhetoric as the art of persuasive speaking placed the power of language into the public sphere and provided the users of the social order, the citizens, power that previously had been in the hands of only the most privileged classes.[22] In concert with the

---

22. This does not mean to imply, of course, that Greek society suddenly became nonhierarchical. On the contrary, there was still a very limited number of

advent of rhetoric, Vernant points to the openness of the social order as the second great feature of the *polis*. The secrets of the priests or monarchs were now open to public view, with the consequence that what were once "truths" understood by only a few were now available for speculation or debate. "Knowledge, values, and mental techniques, in becoming elements of a common culture, were themselves brought to public view and submitted to criticism and controversy" (p. 51).

The third element of the polis in many ways combines the other two elements and is one full of tensions, especially when interpreted through our postmodern context. "Those who made up the city, however different in origin, rank, and function, appeared somehow to be "like" one another. This "likeness" laid the foundation for the unity of the *polis*. . . . In the sixth century this image of the human world was precisely expressed in the concept of *isonomia*—that is, the equal participation of all citizens in the exercise of power" (pp. 60–61). The paradoxical tension of this egalitarian vision of a social order rests in the difficulty of creating "likeness." On the one hand, to be like one another describes a system of equality and shared exchange of power. Hierarchy is refigured as a horizontal plane where everyone shares across a level space, instead of a vertical plane where decisions and power move from top to bottom (usually) or vice versa (rarely). On the other hand, "likeness" posits a scenario that could very well disregard individual differences, as these differences would be subsumed by the common look of the social order. A contemporary example of this difficulty is found in our debates over "color blindness" regarding racial equality. To be "color blind" racially, some argue, means that all races can be seen as equal. Others counter, however, that the melting pot-like metaphor of "color blindness" is actually a definition of equality created by those already in power who want to maintain the status quo. In other words, once social differences are combined into a color blind "likeness," then we have erased the distinguishing differences of many cultures and forged a single "look" that, like the physical properties of light, results in one color—white.

Without a doubt, a user-centered theory must confront the dilemma of "likeness" if a goal of the theory is, in part, to aid in the achievement of a more egalitarian technological order. The first two elements of the polis

---

citizens allowed to participate in the debates at all. Women, non-landowners, slaves, and others were denied access, in general, to these new powers of language in thep-ublic sphere. The events associated with rhetoric's birth were only the beginning in Western culture of a recognition of broad-based power and knowledge—something we continue to struggle with to the present day, especially in the context of literacy studies.

described earlier—those of the preeminence of rhetoric and the openness of the social order— are potentially strong elements of a sociotechnological order that would welcome the knowledge of users. These elements of the polis, though, are hinged to the paradox of "likeness," and we are obliged to confront this complication while developing theories of user knowledge so that they, too, are not blind to the invisible hierarchies that already subvert the user in the sociotechnological order.

No answers are complete, nor are they immutable. However, we can begin to confront the problem of constructing a polis truly open to users, one that is egalitarian in its best sense, if we return to the problem of user knowledge and consider it as a *knowledge of action.* One way to couch this knowledge as action is to define it as conduct, or how people *know* in the context of conducting human affairs. In the realm of technical communication, Carolyn Miller addresses the problem of social conduct when she explains that technical communication pedagogy should move beyond its preoccupation with merely constructing texts as its end and instead should rethink texts as a means to an end: an end aimed at the common good.

> Understanding practical rhetoric as a matter of conduct rather than as production, as a matter of arguing in a prudent way toward the good of the community rather than of constructing texts, should provide some new perspectives for teachers of technical writing and developers of courses and programs in technical communication. (*Practical,* p. 23)

Similarly, user-centered approaches should rethink the user as being an active participant in the social order that designs, develops, and implements technologies. Users as producers have the knowledge to play an important role in the making of technologies; users as practitioners actually use the technologies and thus have a knowledge of the technologies in action; users as citizens carry user knowledge into an arena of sociotechnological decision making: the arena of the polis, or, if you prefer *politics.*

In the realm of the polis, users are both responded to by, and are responsible for, the technological order. Technological decision making in this new order is recast as a two-way enterprise where users are invited to participate, but because they are members of the polis, they too must reciprocate responsibly toward the "good of the community" because they are citizens of the community. Before we end this discussion of user knowledge, let us look at one last scenario to help cement the concept of users as citizens:

By the early 1980s it had become clear to the Seattle highway department that traffic congestion, particularly during the morning and evening rush hours, had reached crisis proportions. Just like any large urban area experiencing fast-paced growth, the greater Seattle area had finally reached a point where some serious studies of traffic flow were needed. To that end, the traffic commission looked south to Los Angeles for answers. Surely, the commission thought, an area like Los Angeles could teach us much about studying traffic problems.

Consequently, the traffic engineers of Seattle imported a number of techniques from Los Angeles to study the traffic flow problems of Seattle. Most of these methods were based upon time–distance formulae and statistics describing the amount of traffic flow at given times of day. As you might imagine, the Seattle engineers collected a myriad of data from these techniques. This data was put to use by determining, for instance, which roads to reroute at what times of the day, which bridges to open or close at certain times, and which bypasses to rebuild or expand. After the expenditure of much time and money, however, the Seattle engineers came to a startling conclusion—no differences in traffic flow were occurring. The traffic jams were worse than ever. The amount of traffic since they began their study had increased to some extent, but their figures demonstrated that the flow should have been corrected more than it was.

About the same time that the engineers were having problems, a team of technical communicators from the University of Washington proposed to study the traffic problem from a different perspective. Instead of studying the traffic, they proposed to study the people who were driving the cars. In other words, they proposed to analyze the preferences and habits of the drivers to establish patterns of behavior, or what we might call patterns of use. Their study, based on surveys, interviews, focus groups, and observation, collected perhaps as much data as the previous studies by the engineers but this new (and for the most part qualitative) data came to a much different conclusion.

Most pointedly, this data of people (as opposed to artifacts) showed that the traffic problems might be best addressed by giving people daily information about traffic patterns so they could make choices about which route to take on a given trip. So, techniques of providing information to drivers as they prepared for work in the morning, or as they prepared to come home at night, were presented through an array of media—the television, telephone, personal computer, fax machine—and in a fashion that fit users' needs and their changing contexts. The technical communicators also made another interesting determination: they concluded that they would never get all of the drivers to alter their habits. Consequently, they aimed their new techniques at those drivers who would most readily change their routes voluntarily according to the traffic flow problems of the day. This way, the communicators decided, they could effectively solve a significant portion of the problem as opposed to risking ultimate failure by trying to get everyone to change his or her daily habits. As the old saying goes, "You can bring a horse to the trough, but you can't always make him drink."[23]

In this rich scenario we have present a number of the elements of the rhetorical complex. The designers, users, and technological systems were integrated in one effort to design an effective system for use by a variety of citizens; the members of the rhetorical situation negotiated which solutions would be best in the given situation; the constraints of the technological system were taken into account; the social environment was taken into account as well.

In terms of the user as citizen, however, the point is made most strongly: the users are represented as being an important force in the design of the system (the highway information system) because they are asked to help determine the best solutions to the problem. Although it is not explained in depth by the scenario, the drivers were given choices regarding how they would like to receive the information, and at what times it would be most useful to them. The drivers also were valued in the sense that some of the participants, those who agreed to be active in the project, were rewarded by design decisions that played in their favor. Put another way, those who acted as responsible, active citizens reaped the benefits of involvement.

Varieties of user participation, however, need not be limited to examples like the Seattle drivers. Users can participate in technology through forms of "noninvolvement," too, where choices must be made about what direction to take in technological development. For instance, many of us are now advocates and practitioners of recycling. We recycle everything from bottles to boxes; recycling is a worthy activity, there is no doubt. Are there "down sides" to recycling, however, that could be altered by user practices? Take the lowly plastic soda bottle, for example. We now recycle many

---

23. This scenario is drawn from the work of several researchers in the Department of Technical Communication at the University of Washington, Seattle. The scenario depicted was drawn from my notes of a talk given by Mark Haselkorn at the 1989 SIGDOC Conference in Pittsburgh, Penn. For specific information on this project, see Connie Miller, Jan Spyridakis, and Mark Haselkorn's "A Development Tool for Advanced Traveler Information Systems Screen Design," Internal Report Number FH–40.

of these bottles, and the products that come from this effort have expanded to more than just new soda bottles. The plastic is used to make carpets, carpet padding, garbage cans, picnic tables, and packing material, among other things. Reuse of the bottles is certainly preferable to throwing them into a landfill, or worse, tossing them into a ditch by the side of the road. But is this the only solution? By that I mean, do we question the products that are contained in the bottles, or the fact that the products we consume are part of a complex system of manufacture, packaging, transport, and advertising that uses vast amounts of energy and natural resources? What would happen if instead of always buying soda in small, recyclable bottles we became advocates of consumer "noninvolvement?" What if we made our own soda, at least part of the time, thus *lessening* our involvement in the consumer practice of buying the ready-made product? True, we might have problems carbonating the drinks if we lack the proper equipment, but then again I am sure our teeth would reap some benefit from the reduction of carbonic acid in our diet. We also might miss the flavor of the cola's secret formula, but we might just discover a better one.

Certainly there are numerous problems with such suggestions. Commerce, industry, agriculture—these are all enterprises dependent upon our current system of production and consumption. The more we consume, the more we must produce, and, we are told, the more jobs there will be. We have become a culture that wears its labels on the outside of its clothing, and we seem quite comfortable with that display of end-use consumption. We seem comfortable only driving, not building and driving. Does user knowledge really have anything to do with these large problems? Can a refigured sense of what it means to *use* really create changes in our basic culture? We might try, and as a result we could very possibly *reinvent,* users' ways of knowing. People as producers, people as participants, not as idiots. Imagine.

CHAPTER 4

Human Factors and the *Tech(no)logical*

Putting User-Centered Design into Perspective

A large wristwatch manufacturer has just had a highly successful year with its newest digital timepiece. Suddenly, though, a significant number of the pieces are being returned to the manufacturer as "defective." Upon inspection of the timepieces, the manufacturer discovers that the watches work perfectly—no defects whatsoever could be found. Puzzled, the company invites a team of human factors consultants to investigate. What did their investigations reveal? The consultants concurred that the watches were indeed in fine working order. However, they also discovered that they were being returned at two distinct times of the year—in the spring and fall, when the time changes.

—adapted from John Sedgwick, *The Complexity Problem*

A leading manufacturer of fax machines spent a great deal of money and development time enhancing one of its leading fax products. The result was the addition of three key features that would allow users to complete tasks that would make the units more appealing. Some time after the introduction of the new machines on to the market, the company surveyed the purchasers of the enhanced machines to see how well they liked the three new features. The result: 95% had never used them.

—adapted from Herbig and Kramer,
The Phenomenon of Innovation Overload

The aforementioned scenarios illustrate a fundamental difficulty associated with technological use. This breakdown between humans and technological artifacts is called several names, such as "the complexity problem" (Sedgwick 1993), "innovation overload" (Herbig and Kramer 1992), or, to paraphrase Thomas Landauer (1995), "the trouble with technology." The result of this mismatch between people and technology, however, is usually the same: the artifact is either misused or not used at all. Some may argue (for example, from an "idiot-proof" perspective) that this apparent mismatch between the machine and the human is the result of an untrained or a lazy human who just does not understand, or does not care to take the time to understand the complexity of the technological system. This point of view, in fact, is predominant in Western culture (and maybe more globally). It is a sort of "the few, the proud" mentality of technological use that assumes everyone is situated in the same time and space, with the same interests and values.

Technological systems have been routinely designed under the aegis of two basic premises: 1) that systems experts are the best equipped to make design decisions, and 2) that the best systems are designed to reflect rational decision-making processes. The breakdown of this dominant perspective of systems design is apparent in the two preceding scenarios. In the first, the experts designed an elegant, perfectly ordered electronic timepiece, but it failed in practice because the "logic" of the system was not visible to the user. In the second, the system experts enhanced a facsimile machine, thus making it more appealing and, hopefully, more useful. In practice, however, there was a breakdown because the users did not care to use the enhancements. I am sure that the designers felt the enhancements were "logical," but the users obviously did not agree. To the users, these devices were, so to speak, "tech(no)logical."

A user-centered point of view, however, argues that such breakdowns are not the fault of the user but instead are the result of technological design processes that fail to involve users during the design and development of the artifact. Most current technology design processes are based upon rational models of human behavior that attempt to predict the logical series of actions the potential user will follow in order to use the artifact. As the two scenarios illustrate, however, many situational aspects of technological use are ignored by this "logical fallacy" of expected user outcomes. Technology designers, for instance, may find it interesting to hide innumerable features behind simple interfaces like digital wristwatches. But if the user cannot see them, or if the user does not readily see how to access these features (do you remember the last time you tried to set the alarm on the digital clock radio at an unfamiliar motel just before you wanted to go to sleep?), then the artifact is failing, regardless of how sleek or perfectly ordered it may be. Put another way, even though the artifacts in the aforementioned scenarios were systemically or functionally perfect, they failed (or at least were not used to their expected potential), because the "perfectly ordered system" was not made apparent to the users.

The study of such disjunctions between users and technological systems has been, for the better part of the present century, the domain of human factors: a large and complex discipline that involves specialists from fields as diverse as psychology, anthropology, computer science, statistics, communications, and physiology. User-centered design has basically been the result of research done in and around the general field of human factors. Consequently, many attempts to apply user-centered concepts to technical communication have been drawn from human factors. At first blush, the idea of borrowing from human factors research appears to be quite sensible. Indeed, human factors research can be quite informative when it comes to simplifying the sometimes overwhelming complexities of human–technology interactions. At the same time, however, the use of human factors research by technical communicators can be wrongheaded, or at least misguided, because a tendency in borrowing research from "foreign" fields is to reduce that research to a dangerously simple level. In addition, such borrowing can cause the borrowers to ignore their own strengths and instead opt for solutions based on what appears to be reasonable advice from another discipline.

In this chapter I want to confront the problem of borrowing from human factors research as we develop a technical communication conception of a user-centered view of technology. In particular, I have two purposes. The first is to investigate the rational proclivity that is historically embedded in human factors research. This focus on the rational actions of human users, I will argue, actually perpetuates the development of system-centered technologies that merely give the appearance of focusing on the needs of humans (i.e., the words "human," "user," or "man" are used in conjunction with "technology," "machine," or "computer" to define different research areas of the field). To accomplish this I will begin by working historically to investigate some defining moments in the history of human factors research during the present century. I will end this historical investigation by highlighting some areas of current research in human factors that are involved in refiguring the rational approach by addressing what I call the "tech(no)logical": an essential paradox that results from the interaction between humans and machines.

The Spectrum of Human Factors

Even though technical communication is a relative newcomer to the discipline of human factors, there has been considerable research indicating its importance. For example, several researchers have investigated user documentation through the lens of human factors (Felker 1981; Haselkorn 1988; Sullivan 1989; Oborne 1989; Carroll 1990, Johnson 1994), and studies of text readability have been conducted for a number of years (Tinker 1963; Duffy 1985). In the pages that follow, we will survey the disciplinary diversity that constitutes this widespead field for the purpose of conceptualizing its relevance to the present argument of user-centered technology and its relation to technical communication.

There are many divisions in the field of human factors, and one element that defines these divisions is the research aim of each particular subfield. The research aims are vastly different and how the research findings of these different fields impinge on the work of technical communicators is important to consider (see Figure 4.1).

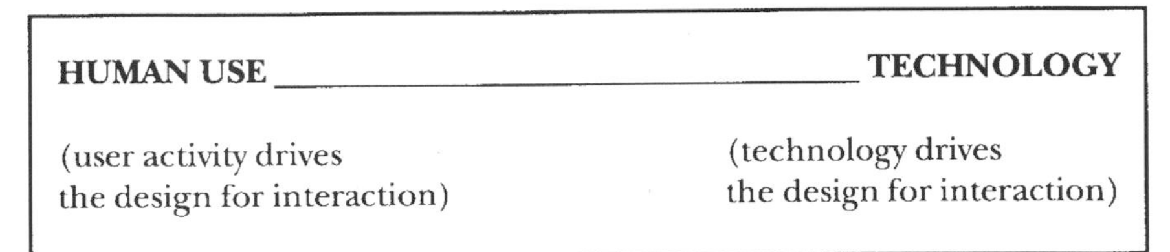

Fig. 4.1.
The Human Factors Spectrum

On the far right-hand end of the spectrum are the aims of technology. Essentially, these aims are concerned with how to develop technology that is most efficient, in terms of quality, cost management, and time. This research is driven by the concerns of the technology first—the idea of a "user" is a lip-service concept that justifies the need for the technology. Toward this end of the spectrum are those areas of human factors that attempt to design for the user, but are not user centered. Ergonomics, efficiency studies, and human factors engineering assume little or no user input at the early design stages.

In other words, these areas of study wish to make systems more usable, but they still base their research on models of the system that come from a system designer's viewpoint. The user is merely represented in the designer's mind, and the theories assume or require that the designer be sensitive enough to understand the needs of the user. These system-centered theories also place the designer in a highly dominant role that virtually excludes the user from any direct influence on systems development, except in cases where a user-sensitive designer creates an accurate fiction of the intended user—cases that are extremely rare in the current paradigm of system-centered technological development.

The far left-hand end of the spectrum is occupied by concerns for the active human use of technology. Instead of basing design and development decisions on rationally based theoretical models, researchers on this end of the spectrum wish to build rich understandings of actual users working in real (or at least simulated) situations. The research here, then, is driven by understanding how people use technology within the constraints of the situations of everyday activity and as such is a practice-driven approach to theory building situated most often in qualitatively oriented empirical study. Such approaches resist the "idiot-proofing" mentality of the system-centered view (Bannon 1985) that assumes systems are fool proof and that errors or miscalculations can be attributed to the user.

In fact, some radical approaches to human activity or user-centered design take the extreme view that "human error" in terms of user error does not really exist. Instead, error is seen as a result of system design that neglected the user in the design phase. One of the most tragic examples of the failure of a system to meet user needs is that of the British bomber squadrons in World War II (see Norman 1983). The British were losing a much higher percentage of crews than were the Americans, even though the number of planes shot down was about the same for both countries. Through functional testing of the aircraft, it was found that the British bombers were equal to the American bombers, and it was determined that the crews were equally well trained in flying techniques. It was hypothesized that the British crews might not have been trained as well in emergency measures, or that they were not carrying out their procedures of escape in the correct fashion. In other words, it was suggested that the British bomber crews were committing a greater frequency of human error. It was discovered, however, that the escape hatches in the British bombers were smaller than those of the American planes. That is, the British bomber's escape hatch could accommodate an average-sized man, but not a man with a parachute strapped to his back. Certainly some simple on-ground simulations of escape procedures would have rectified this tragic design flaw.

Clearly, the human use perspective argues that the ineffectiveness of systems lies in the miscalculations and poor planning of the designers, the changes in environment, or alterations in user needs. For instance, human use philosophy advocates the assessment of user needs *before* system development is even initiated, and that the system designers should develop models of systems in accordance with users' needs (Bødker 1991). The early assessment of user needs turns much of the designing planning process around and consequently directs the development of systems toward a collaborative effort of designers and users.

Truly placing the user at the center of technological development means understanding what constitutes the user's situation, and this, in turn, means defining the scope of this situation. For example, in the case of instructional materials, will the user be using the documentation to learn about the artifact, or will the user be going to the documentation merely for a quick reminder of a temporarily forgotten function? Is the user working individually, or is there a collaborative situation framing the context of use? Before we can address these or other questions, though, let us turn to a selective history of human factors research to see its constituent parts, and which areas of this broad field are most relevant to technical communication.

Human Engineering or Engineering the Human?: The Beginnings of Human Factors Research

Some researchers take quite a broad view of the history of human factors, and they go as far as placing its beginnings at the earliest moments of human development when humans first began to develop tools from organic and inorganic matter and use them (Christensen 1962). It is more appropriate for our purposes, however, to begin where the study of humans and technology started to develop as a discipline—that is, when researchers consciously began to study how humans interacted with technology and how the findings of their research could be used to improve the quality of those interactions.

From the very beginning, human factors research has been driven more by a concern for technological and economic development than by a concern for the user of the technology. For example, in 1832, Charles Babbage wrote his book *Economy of Machinery and Manufacture,* where he described methods for alleviating the physical demands placed upon industrial workers by using the "division of labor" as developed by Adam Smith in *The Wealth of Nations*—a concept that would have tremendous ramifications in twentieth-century industrial development (Burgess 1986). While division of labor did accelerate the level of production, it also relegated the workers to continuously repeated tasks that often led to boredom, fatigue, and other problems related to over-specialization. The ramifications of specialized, repetitive tasks carried out by workers is still evident today, as illustrated in the frequent reports of "repetitive motion syndrome" suffered by office personnel and assembly line workers.

Economic development and worker specialization continued to be emphasized in the latter half of the nineteenth and early decades of the twentieth centuries through the work of Frederick W. Taylor. Taylor designed and implemented empirical studies to investigate problems such as the design of shovels and the techniques of manual loading of steel ingots. He describes the four basic principles of this empirical method in his book *The Principles of Scientific Management:* 1) scientifically justify each element of a worker's job in terms of its contribution to efficiency, 2) scientifically select the workers and train them, 3) ensure that the job is done as prescribed, and 4) have active management participation in the job.

His methods were based on a rationalistic view of how to best match "an organization's means to profitable ends" (Adams 1989, 378) through the scientific management of workers. A most telling description of Taylor's method is given by Jack A. Adams in *Human Factors Engineering*, and the following example succinctly describes how methods like Taylor's were designed to engineer the human into the system with little consideration given to the human operator.

Taylor had been commissioned by Bethlehem Steel to study the efficiency of workers who loaded steel ingots onto flatbed railroad cars. Adams describes Taylor's method as follows:

> Each ingot weighed 92 pounds, and each man on average loaded 12.5 tons per day. Taylor studied the workers and their work behavior and concluded that it was possible to load 47 tons per day. To prove it, Taylor chose a worker who was strong (the worker ran home every night after carrying iron ingots all day), and who was not too intelligent (Taylor believed that bright workers found repetitive work monotonous and boring), and who would do exactly as he was told (so that the new steps of the restructured job would be followed). The man started to work, being told when to pick up an ingot, when to walk with it, and when to sit down and rest. At the end of the day he had moved 47.5 tons, and he continued to do so thereafter. His increase in productivity earned him a pay increase of 61 percent (from $1.15 to $1.85 per day in 1898). Management could easily see that the increase in wages was more than offset by the increase in productivity, and they were impressed. Taylor selected other workers and trained them in the same way. (7)

Taylor's four principles of scientific management are clearly demonstrated by this example, as is the emphasis on efficiency and increased production. The goal of 47 tons per day will increase the efficiency of production; the worker subject is chosen scientifically to meet the goals of the experiment; the work is prescribed explicitly; and management is active in the job (i.e., management oversees the job). The efficiency of the production system and the prospects for economic success are increased, but the human in this research method is of little consequence beyond his or her use as an extension of the technology and the greater economic system that the technology supports.

It is telling, then, that several human factors researchers point to Taylor's work as the beginning of human factors research (Chapanis 1965;

Christensen 1962; Burgess 1986; Adams 1989). Taylor's early research does obviously involve the study of humans and the activities of work, but the focus is on the ends of productivity and efficiency from the standpoint of management. Such research is the precursor of the system-centered design philosophy that drives much human factors research to the present day.

Computers and End Users Enter the Scene

As the field of human factors research entered the 1970s and 1980s it brought with it a new interest in the use of systems within socially situated environments, and one result was a concentrated focus on designing for the users of systems, as opposed to designing merely for the efficiency of the system itself (Suchman 1987; Brown and Newman 1985). Prior to these two decades, human factors research most often concerned itself with technology that had already been put into use. Put another way, human factors specialists took existing systems and then taught users how to use them. This traditional system-centered viewpoint assumed that design problems involving users are often insurmountable and that the solution to misuse is to teach a user how to deal with design shortcomings.

The idea that the user must be integral to the design process has gained considerable momentum over the past fifteen years, but to view this as a singular movement in the field of human factors would be a mistake. For example, the aforementioned interest in the overall environment of human use often has been mere lip service during the past several decades. Although human factors researchers have begun to acknowledge the problems of environment, they often do not practice this philosophy in their research. The choice of how writers will implement human factors into a theory of computer documentation, for example, must be a decision that is made with a conscious awareness of how users and systems are viewed by alternate perspectives in the field. Therefore, an example of one such debate is in order.

Hard Science versus Soft Science in Human–Computer Interaction

As was already mentioned, human factors engineers traditionally tested existing systems on users to ascertain whether the system could be effectively used. Attempts to incorporate even minimal amounts of information from users into the design process were almost unheard of, a reason being that the potential users of most systems were experts, or at least fairly highly trained individuals. The advent of networked computer systems, and later the microcomputer, changed this. No longer was high technology only within the realm of the scientist, engineer, or astronaut. Complex systems existed within the everyday world, and millions of uninitiated users were suddenly grappling with the use of sophisticated machines.

Human–computer interaction (HCI) specialists, however, are divided on just how system development should be altered to meet the new demands brought about by this expansion of user populations. On the one hand, some researchers (termed *hard science* advocates in the upcoming discussion) still cling to a system-centered view that is more interested in the internal specifications and functions of the computer than with the user's psychological or social interactions. These system-centered researchers perceive the computer as a human cognitive modeling device rather than a tool of use. Their intent is to understand the scientific principles underlying cognitive processes and how these universal processes can help drive the design of computer systems.

On the other hand, there is a school of researchers (termed *soft science* advocates in the upcoming discussion) who attempt to place the user at the center of the design problem. They are more concerned with the human as a user of systems and therefore focus their research on the problems associated with the psychology of everyday activities: activities that may or may not have any rational, logical, or consistently verifiable processes. Thus they advocate system design processes that employ user input as something that can help create more usable systems designed for everyday practices. The first group of researchers—the hard science group—is more interested in deriving more or less steadfast, universal scientific principles upon which to base design decisions, while the second group—the soft science researchers—focuses on the situated user as the central criterion of design. The differences between these two schools are methodological, but there are philosophical contrasts too. To illuminate the nature of the differences between these two schools, let us focus on a recent debate by two pairs of researchers who hold opposing viewpoints in this matter.

In a series of three articles appearing in *Human–Computer Interaction*, the teams of Allen Newell and Stuart Card (1985) on the one hand, and John Carroll and Robert Campbell (1986) on the other, argue whether human–computer interaction is a "hard" or "soft" science. Essentially, Newell and Card wish for HCI to "harden up" in its methodology, or they fear the impact of psychological science on the field will be minimal at best and nonexistent at worst. Although they view psychology as being an important part of HCI development, they question whether the qualitative methodology of soft science techniques will be accepted by the more quantitative fields of computer science, engineering, and so on. To clarify their position, Newell and Card employ the analogy of Gresham's law: a law of economics that was formulated in the sixteenth century by Sir Thomas Gresham to "describe the fact that newly minted coins tended to disappear from circulation, leaving only the worn ones" (211). Newell and Card condense this law into the aphorism, "Bad money drives out good," and paraphrase it into the province of HCI to say, "Hard science drives out soft" (211). They define a "hard science" as "science that is mathematical or otherwise technical," and they contend that soft science methodologies of psychology can only be of benefit to the study of human–computer interactions if psychology will adopt the "harder" methods in its applications to HCI research. In fact, Newell and Card go so far as to state that they "see no other way of defeating Gresham's law. Hardening the science is certainly not the easy path. It counsels getting the human concerns into the interface by just producing better science" (222).

Carroll and Campbell, however, argue that HCI should not confine itself to such "strict methodological strictures." They contend that Newell and Card are too narrow in their approach to interface design and that the use of Gresham's law typifies the narrowness of their philosophy. Carroll and Campbell reparaphrase Gresham's law to read "overvalued money drives out undervalued money," (228), and they contend that "the proper course of action . . . is to redress distortions in value, not to accommodate them; to protect and develop conceptually deep science, not to abandon it" (228). Carroll and Campbell go on to question the tenet of hard science, and they remind us that in the history of science, qualitative methodology laid the groundwork for quantitative methods (244). They accuse Newell and Card of "reviving positivism" (246), and they contend that such an approach to human interaction with computers ignores the real problems people face when using computers. Put another way, Carroll and Campbell see Newell and Card's approach as one that is interested in a search for universal truths that can be applied to the understanding of computers and the humans who use them. Carroll and Campbell, in deference to this view, seek methods that will enable us to develop more human driven explanatory understandings (as opposed to technologically driven descriptive understandings) of the psychology of human–computer interaction.

It can be argued, however, that quantitative empirical evidence and applications based on the findings of such research are needed in order to build a theory of users, therefore Newell and Card might offer the best solution to developing an acceptable theory. By using hard, quantified, and verifiable data, we might be able to construct formulae that would tell us what to do when designing interfaces or writing documentation. Unfortunately, the choice is not that simple.

To begin with, the approach represented by Newell and Card places the system ahead of the user. This research is interested in the computational modeling of the mind as an information processing machine (Boden 1988). The focus is on the mimicking of human thought processes and is based on Newell and Simon's work regarding decision making (see Newell and Simon 1972), and more recently on Newell's own research into the GOMS (goals, operators, methods, and selection rules) family of models. Such research is driven by concerns of artificial intelligence and not by the human use of systems. Artificial intelligence research of this sort is important to the development of systems and to a large extent the understanding of human thought processes—especially those processes that fall under the aegis of bounded rationality (Simon 1970, 1977). However, the aim of this research is constrained to uncovering and understanding general principles of human cognition to predict behavior that can then be imitated by a computer. Admittedly, artificial intelligence research of this sort is not confined to understanding the internal workings of the human mind, nor is it limited to investigations of the effect of external environments upon systems, as some critics have argued (Boden 1988). It is, instead, research that attempts to model general principles, and the specific actions of everyday experience are overlooked, or at least not prominent. Thus the computer system, as a system that can adapt itself to the rational solving of problems within a given problem space, is the priority here. The user, as an unpredictable actor in unpredictable situations, is not of concern.

The unpredictability of the user brings us to another dispute argued by Carroll and Campbell and Newell and Card. Human nature is too uncertain for formulae or algorithms that attempt to define universally the complex study of human–technology interaction. This is made an issue when Newell and Card discuss what they label as "descriptive models" in task analysis. As Newell and Card explain, "Theories of human–computer interaction ultimately need to be able to give accounts based on explicit representation for both the concepts of the theory and for the operations—as opposed to judgment, experience, or empirical evaluation. . . . " (Straightening Out, p. 260). These "descriptive models," Carroll and Campbell argue, are not concerned with the actual tasks that users carry out on the computer. Instead, Carroll and Campbell define an "explanatory" approach that "takes as its starting point the need to understand the real problems involved in providing better computer tools for people to use." It may be true that for certain research agendas the descriptive models limit the scope and therefore the variables for possible error, but this is too narrow for an adequate understanding of real users.

This points out the adherence by Newell and Card to traditional human factors research where systems are to be used by or modeled on expert users, but even more it demonstrates the interest that these two researchers have in universal scientific principles of the "ideal." Once again, Carroll and Campbell point this out when they say that Newell and

Card's "model is a model of ideal experts, not of real experts or real novices" (p. 241).

As we can see from these historical selections, the drive to use methods based upon rationality and scientific rigor has dominated the field of human factors research for virtually its entire existence. The Taylorist principles of scientific management attempted to model the human as a rational being who, if given the proper rules to follow, will invariably be a more effective (and thus more profit generating) worker. Many of these same principles are still used today, although as Carroll and Campbell demonstrate, there are significant challenges to such rational approaches underway.

As technical communicators begin to apply human factors research to technology development, clear understandings of which approach to use will become quite important. For example, theories of computer documentation and its audiences are in a very early stage of development—much newer than the work done on the rationally driven theories of computational modeling of the mind—and need to have a good deal of qualitative work done before quantitative validations can be used to accurately test documentation. In addition, the study of writing processes and those users who use the resulting texts is not a hard science. Instead it is a human/social science that gains understanding from the observation of human nature. From these observations of writers and users within the actual contexts of use, we can develop the strategies that technical communicators can use in everyday writing tasks. We need flexibility in our methods to meet the demands of changing technology and fluctuating contexts. The methods of hard science will not be appropriate at such early stages of understanding, because hard science tends to be concerned with the "ideal," and writers operate in the less stable world of the contingent.

"Soft science" approaches to human–computer interaction research have proliferated over the past five years, and the impact of these methods is beginning to be felt quite strongly in the design and writing of computer user documents. I would like to conclude by describing several subareas of research in human–computer interaction studies that hold great promise for technical communicators. To begin, I will overview minimalist design and usability. I will conclude the chapter with a similar overview of some European-influenced design approaches, namely human activity design and participatory design.

Studying Users Through Usability and Minimalism

Usability is a term that comes from computer interface design, and, in a very loose sense, it involves the ease with which a user (or users) can learn about or use a computer system. Interest in the usability of written documents, though, is a relatively new concept. During World War II, the use of instructional material served as a training aid for military personnel to learn about the safe and correct use of military technology, but the function of the documents as a direct aid in the use of these complex technologies was not a topic of concern. In other words, radar operators or battleship gunners were given instructional documents in their training, but how the documents were used during the interaction with the technology was not studied until the 1970s. In the 1970s, researchers on both sides of the Atlantic began investigating the importance of documents to the human use of technology. In Britain, this research was spearheaded by James Hartley (1985) and Patricia Wright (1983), and the focus was on the visual design of information and the cognitive processes involved with the reading and use of functional documents.

In the United States, there were two primary centers for document usability research—the American Institutes of Research (AIR) and the military. The AIR grew out of the plain language movement, and their research helped develop the "process model of document design": a model based on the prewrite/write/revise process model, but which also incorporates visual design, system constraints, and document production processes (AIR 1982). The military continued their interest in the development of procedural documents begun in World War II, but their research became more concerned with the use of documents in actual situations. From this research came the important distinction of functional reading tasks as either "reading to learn" or "reading to do" (Sticht, et al. 1977; 1985)—a distinction that has proven helpful in the development of current functional text theory and that also is useful in discussions of computer documentation theory and pedagogy (Redish 1988).

In the 1980s document usability research became increasingly interested in testing documents to determine such things as the accuracy, completeness, and usefulness of texts. Usability testing of computer documents has a number of different definitions, but it has been characterized as coming from two different viewpoints—testing and evaluation (Sullivan 1989). The testing view is the more narrow definition—where usability is concerned with the testing of documents after they have been drafted and often carried out under laboratory-type conditions. A broader view of usability, as evaluation, advocates usability as an iterative process that contributes to early pre-draft evaluation of document needs, as well as testing that takes place later in the document development cycle.

The growth of documentation usability over the past decade has helped produce much data concerning the nature of users and how they use written documents and computer systems. One benefit of this data collection has been the beginning of theory building regarding users and the translation of these theories to documentation writing. For example, in the reality of the nonacademic workplace, the writers of documentation often are separated from the usability specialists. Consequently, writers continue to serve the purpose of editors who take the findings of usability specialists and revise the documents accordingly. By using the findings of usability research to build strategies for documentation writers, we gain knowledge about users and put this toward the practical purposes of giving documentation writers strategies for developing more usable documents.

Mary Dieli of Microsoft Corporation and John Carroll of IBM are two of the more visible practitioners of bringing usability research to documentation writers in the workplace. Dieli has developed what she calls "revision filters," or methods for document evaluation based upon studies of computer user behavior (1988a). These "filters" are essentially heuristics that writers can use to gain "an accurate user perspective about a document as it is being written, rather than after it is written" (p. 151). In addition, Dieli (1988b) has proposed problem-solving strategies to aid in the design of user testing so writers can have a flexible, yet focused, approach to usability projects within the constraints of the workplace.

Carroll's main contribution is the "minimal manual" (1987a, 1990). Using thinking aloud protocols, Carroll and his associates explored the problem of how users learn during the activity of computer usage. They concluded that novice users become easily frustrated during the process of learning a computer system and that the tutorial documents that were written to help the learning process were woefully inadequate because these documents did not support the active learning styles of the novice computer users. In general, Carroll and his colleagues determined that the current tutorials were based upon the assumption that users will read the tutorials. Usability research, however, provided evidence that this was not the case (see also Wright 1988). Novice users want to become involved with the use of the computer immediately, and reading becomes a hindrance because it disengages the user from the activity of computer usage. Thus, Carroll drew upon evidence about novice users from usability research to develop guidelines for writing manuals of minimal size that avoid getting in the way of the computer learner by providing adequate, but not extensive, information.

Designing for Human Activity and User Participation

As the previous example of minimalist design points out, the active state of users should be a primary concern of documentation writers. For documentation to be effective, it must be designed to fit into the active situation of the user, or it will never be used. In the last five years, two "schools" of interface designers (human activity designers and participatory designers) have developed that are concerned with the design of computer systems for the active user who uses computers in socially constrained situations. Although little if any of this research has been applied to documentation development, I would like to briefly overview these two schools of designers, as they could very possibly set the stage for important research in future documentation theory and practice.

Human activity designers and participatory designers describe schools of computer interface design interested in developing computer systems that meet the needs of users involved in what one researcher calls "purposeful human work" (Bødker 1989, 171). A main characteristic of these researchers is that they pursue interface designs that are not based upon rational structures, but instead reflect the often irrational and unpredictable flux of human nature.

> Their work is part of a new and evolving theoretical approach that considers human activity, design, or use of computer applications, not as primarily characterized by rationality, planning, and reflection, but characterized by practice and our ability to act in situations that are more or less familiar to us, wherein reflection is something secondary or post factum. (Bødker 1989, 173)

The aim of these researchers is to broaden the perspective we have of what computers are and how they are used. To make a connection to the vocabulary of rhetoric and composition studies, human activity and participatory designers are interested in the social, political, cognitive, and practical facets of computer usage.

In the United States, many of the advocates of such design methods are located at the Xerox Palo Alto Research Center (Xerox PARC), and the chief investigators are interested in developing computer systems that can function in social contexts (Suchman 1987). As Brown and Newman explain in their article "Issues in Cognitive and Social Ergonomics: From Our House to Bauhaus," computer systems must be perceived of as being more than just systems per se, and that "it is the interplay between information systems and social and cultural systems that will make the difference in effective individual and collective use of computational systems" (1985, 382). These same sentiments regarding systems design are echoed by two other researchers, Winograd and Flores, in their book *Understanding Computers and Cognition: A New Foundation for Design.* They take a broad philosophical view of the concept of design in general, and define it as a concept that addresses "the broader question of how a society engenders inventions whose existence in turn alters that society" (1985, 4). From such a perspective, then, come interface designers who share a concern about users as being vital components of computational systems: components that drive and define the makeup of those systems.

In Europe, the counterpart to the American designers is centered in the Scandinavian countries. These designers are highly motivated by socially driven goals, and while they admit that there is no well-defined school or "Scandinavian Approach," they declare that their overall efforts in interface design are aimed at "humanization and democratization as overriding design goals, in keeping with the aim of building an egalitarian society" (Floyd, et al. 1989, 253).

Clearly there are movements in positive directions concerning human factors. Participatory techniques that actually invite users into the design process, egalitarian goals that drive decision making concerning technological innovation and implementation, and the use of qualitative research methods that account for the larger context of use are all directions that should help redefine some of the discriminatory notions we have of users and technological use. As we shall see in the next chapter, however, relevant and up-to-date methodologies are not the complete answer to the user-centered problem. Technology itself, at least as we have come to know it in the twentieth century, may be the most formidable barrier of all.

CHAPTER 5

Sociology, History, and Philosophy

Technological Determinism Along the Disciplinary Divides

John Henry hammerin' on the right-hand side,
Steam drill drivin' on the lef',
John Henry beat that steam drill down, Lawd, Lawd,
But he hammered his fool self to death.

—from *John Henry*, Anonymous, adapted by Alan Lomax

Through the rhetorical complex of technology introduced in chapter 2, I argued that the rhetorical situation involving users, designers, and artifacts should interact in a negotiated manner so that technological development, dissemination, and use are accomplished through an egalitarian process that has its end in the user. The two chapters that followed explored how users and their knowledge of technology have been/can be defined. Chapter 3 examined a potential *user perspective* of technology through a taxonomy of "user knowledge" that describes how users can play an active, participatory role in technology development. Chapter 4 focused on the *designers' point of view* of users by historicizing the history of human factors to show that, for the most part, this discipline has favored designers' goals over the needs of users. In this chapter, we will complete the discussion of these three elements of the rhetorical complex of technology by focusing on the *artifacts and systems* of technology. Taking the issue of technological determinism as the point of significance, we will investigate a perspective where the user is potentially erased from the rhetorical/

technological situation as artifacts and systems become the primary agents of knowledge and cultural development.

Technology, Agency, Society: Sites of Determinist Skepticism

In what some have called "our postmodern world," it is commonly argued that individual humans are no longer the agents of discourse or knowledge, but that agency resides in the social/cultural sphere. This is a way of saying that the ownership of discourse and knowledge is a community activity—that what we know and what we express through language is not entirely our own, but is instead "owned" by us all. Arguments such as this over ownership are not new. As we have already seen in chapter 3, the advent of the *polis* opened the ownership of discourse and knowledge to the populace, thus helping change the course of ancient Greek society by placing decision making in the public sphere. A similar phenomenon occurred in the sixteenth century as knowledge was again opened through the advent of the printed text. As one historian has argued, the very private knowledge of the alchemists was opened to the public sphere through printed technical manuals that described alchemists' techniques of metallurgy. These manuals were then used by commercial mining and metallurgy interests for the purpose of expanding this burgeoning industry in Renaissance northern Europe.[1]

It is significant that these historical moments concerning agency and language were accompanied by technological shifts: shifts that are directly connected to the way language is represented and transmitted through technological media. Once again, rhetoric and the power of language have been key linkages to technology and social formations. In ancient Greece, democracy and the arts of public debate were in part enacted through the new techniques of rhetoric and the emerging technology of writing. The Renaissance, through the technology of print, became a site of widespread information that helped shape our modern notions of textual literacy. Currently, arguments of agency and ownership are being waged in the midst of a communication "revolution" brought about by computer and telecommunication technologies. What is the connection between technology, humans, and agency, between control and ownership? Are these changes linked to, or dependent upon, one another? Is one element of these changes more important or more dominant than the others?

1. See Pamela Long, "Mining, Metallurgy, and the Openness of Knowledge."

Attempting to Define Technological Determinism

Many would refer to this as a problem of determinism, or essential issues of *who* or *what* is the driving force, the controlling power in human affairs. In the context of technology, the issue is simply referred to as *technological determinism,* although some prefer the term *technological autonomy.*[2] Technological determinism is a large subject, and there is no consensus on any one definition. Given such breadth, however, there have been some helpful definitions. One of these is given by John Staudenmaier in his book *Technology's Storytellers*

> It [technological determinism] can best be understood in terms of two major premises and three corollaries. The first premise states that autonomous technology results from a disjunction between efficiency as a norm for judging technical success and all other cultural norms. The second premise argues that technological progress follows a fixed and necessary sequence through modern history. Three corollaries follow. First, the relationship of society to technological change is always adaptation. Second, the historiographic format most congruent with deterministic and progressive technology is the "technological success story." Third, the history of technology is, in fact, an account of the gradual triumph of Western science and technology over all other forms of human praxis. (135–136)

Technological determinism, according to Staudenmaier's definition, places agency in the hands of a broadly defined technology. That is, technology is more than artifacts or systems in this definition and also includes economics, notions of "progress," and narrative constructions of the success of technology in the modern world; many facets of human intellectual and material life, that is, have aided in the construction of technological determinism. In regard to questions of determinism as related to ownership and agency, then, it is not surprising that we find techniques and technological artifacts lurking. Even though we may, in our postmodern context, want to argue that agency resides solely in the social sphere, it is nevertheless difficult to deny the role that artifacts and systems play in the determination of agency. Most pointedly, technologies have a strong and defining influence upon whole complexes of cultural shifts that make the problem of locating agency difficult. It should not surprise us that interpretations of technology as agent are likely to result. Put another way, when technologies accompany cultural changes it often is not clear who or what

---

2. See Winner (*Autonomous Technology*) and Ellul (*The Technological Society*).

is controlling or influencing the change, and the appearance that humans have lost control is a commonly rendered conclusion.

Another way to define determinism is to place it on a spectrum of beliefs about the nature or degree of deterministic possibility. For instance, in the introduction to *Does Technology Drive History?*, Merritt Roe Smith and Leo Marx describe different forms of technological determinism

> . . . as occupying several places along a spectrum between "hard" and "soft" extremes. At the "hard" end of the spectrum, agency (the power to effect change) is imputed to technology itself, or to some of its intrinsic attributes; thus the advance of technology leads to a situation of inescapable necessity. . . . At the other end of the spectrum, the "soft" determinists begin by reminding us that the history of technology is a history of human actions. . . . Instead of treating "technology" *per se* [sic] as the locus of historical agency, the soft determinist locates it in a far more various and complex social, economic, political, and cultural matrix. (Smith and Marx 1994, xii–xiii)

This expanded definition of technological determinism describes the sense that many technology researchers hold regarding the nature of technological agency in the modern world: *It is not whether determinism exists, but rather to what degree it exists that is important.* A visual representation of the spectrum might look like Figure 5.1.

On the right side are scholars who essentially see no hope for humans to control the flow of technological progress. Although there are few at the farthest reaches of this end of the spectrum, there are nevertheless some who have had strong voices, especially during the twentieth century. Moving toward the left side, we have a finely grained spectrum of scholars who represent specializations in areas such as feminist studies, literary studies, the history of technology and science, political science, philosophy, technical communication, and the sociology of science and technology.[3]

---

3. It is beyond the scope of this present discussion to delineate all of the scholars who represent the various positions on this spectrum, and we will be encountering a variety of them during the upcoming discussion of determinism in the sociology, philosophy, and history of technology. A quick reference of those discussing technological determinism would include Helen Fox Keller, Donna Haraway, and Cynthia Cockburn (feminist studies); Langdon Winner, Carl Mitcham, Jacques Ellul, Lewis Mumford, and Ruth Schwartz Cowan (philosophy, political science, and history); Nancy Roundy Blyler, Stephen Katz, Russell Rutter, and Dale Sullivan (technical communication); and Donald Mackenzie, Wiebe

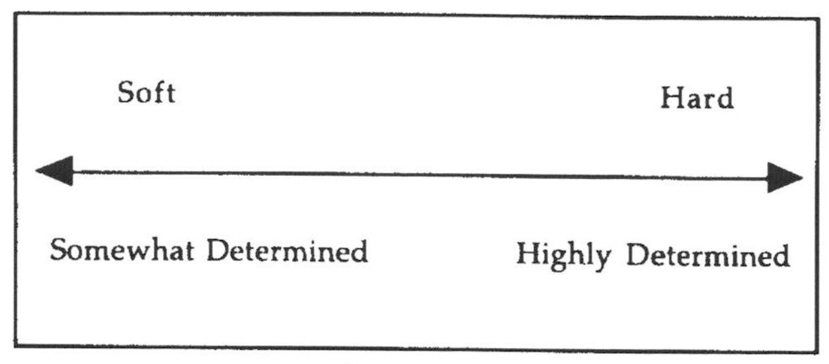

Fig. 5.1.
The Spectrum of Technological Determinism

We will overview representative disciplines that cover the gamut of this spectrum in the upcoming discussions of sociology, history, and philosophy. Thus, this spectrum will be useful to keep in mind as we discuss these researchers and their disciplines in the following pages.

Technology Shaping Culture Shaping Technology

Technology helps *shape*[4] the discursive and material characteristics of cultures. As technologies emerge and are incorporated into a cultural context they alter not just the immediate activity for which they were designed but also have "ripple effects" that shape cultures in defining ways. We often point to those technologies that have had very visible and sweeping cultural effects like the printing press, the steam engine, or the computer. Some mundane technologies, maybe less visible but no less essential, have had equally strong cultural effects and are as influential as any other "shapers" of cultures. The emergence and development of the fireplace chimney in medieval Europe and the accompanying issue of warmth are two illuminating examples.

Well into the ninth century, fireplaces in northern Europe were located in the center of the living quarters—the great halls of the castles or manors of the feudal master—and the smoke from the fire escaped through holes or louvers in the ceiling. In these central rooms, everyone—lords, ladies, servants, and serfs—slept together during the long winter months to take advantage of the heat source. Although often there were

---

Bijker, Judy Wacjman, John Law, and Trevor Pinch (sociology). Their works in regard to determinism can be found in the References Section at the end of the book.

4. See Donald MacKenzie and Judy Wajcman, *The Social Shaping of Technology: A Reader.*

curtained areas to allow for some measure of privacy, the strict social hierarchy of the feudal kingdom was somewhat diffused during the long nights of the winter months. With the advent of the chimney, these sleeping arrangements were drastically altered, and the consequences of this new technology shaped more than just the nighttime activities of Northern Europeans.

> Lord and lady increasingly ate, lived, and slept in withdrawing rooms. As affluence increased, noble residences were redesigned so that rank after rank of the social structure could enjoy the new sense of individuation in its lifestyle. [The chimney] . . . may likewise have fostered the individualism of the later Middle Ages more than all the humanists. Yet a high social price was paid for the new ideal of the idiosyncratic person. As communication between classes decreased, class consciousness and snobbery grew. . . . The chimney is as important as any other single factor in the shift from medieval to Occidental attitudes, and not all of this process was good. (Lynn White Jr., *Medieval*, pp. 271–272)

The social aspect of technology becomes more complex as abstract concepts and the interaction of several technologies are analyzed from a common perspective. For instance, the very idea of *warmth* is equal to chimney technology in affecting our interpretations of the history of human relationships.

The lack of warmth is believed to have had a disastrous effect upon the survival rate of young children in medieval Europe. Children often died before the age of five due to upper respiratory and pulmonary diseases. Consequently, the development of knitting techniques and the corresponding development of the button (circa 1230), helped Europeans create clothes that fit snugly around the small, fragile bodies of young children, thus lessening the risk of these diseases. Beyond the base problem of staying warm, some historians argue that the increased life expectancy of small children led to stronger affection for children because parents and older siblings had less fear of losing young children through death. "It is a safe surmise that the development of knitting, along with functional buttons and heating devices, helped to keep more little children alive, and thus played a large part in fostering modern attitudes toward them" (White 1978, 274). The mix of technology and culture, at least in this instance, constructed to some degree our modern images of children and even notions of family and community.

So where does agency reside in the example of the chimney? What was the main determinant of cultural shift—the craftspeople who designed and built the chimneys, the users who altered and designed the use of the chimneys in their homes, or the artifact itself as it became a part of the culture through widespread use over several hundred years? One could argue that any one or probably all three elements played crucial roles in the social development of the chimney (a position that jibes with the notions of negotiated technological development of the rhetorical complex, set forth in chapter 2). Yet it is difficult to dismiss the power and influence that artifacts have on history, culture, and our lives in general, thus giving rise to skepticism about the issue of human control over technology. As White expressed in his analysis of the medieval chimney, "Not all of this process was good."

Skepticism over the controlling nature of technology is not entirely a modern phenomenon. Even though the ancients valued knowledge of the production and practice of artifacts to some degree, they were skeptical about the inherent powers that new techniques or technologies possessed. The story of Prometheus bringing fire to humankind or the folly of Icarus flying too close to the sun are but two mythological anecdotes for the inherent danger of nature turned to human service through technique and artifact. The ancient philosophers, likewise, projected a concern over the power of technology. Plato's worry that the new technology of writing would destroy memory is probably the most well known (*Phaedrus*), although Aristotle also voiced similar concerns because he believed technology to be at least one step removed from the reality of nature, thus relegating artifice to mere representation of the beauty of nature (*Physics*).

What was skepticism for the ancients though has turned into outright fear of technology's power by moderns. *Frankenstein, 1984, Brave New World*—these are modern visions of the autonomy of technology and its ability to overtake and consume the human soul. These powerful tales of technology out of control—technology going beyond the limits of designers' original intents—describe societies where humans, not technologies, are the "tools" of the culture: cyborglike entities that operate according to a strict, hyperrational logic. Is such a society possible? Can humans become the hapless servants of a technological order, with redemption found only in escape or death? Indeed, are we entrapped in such a situation at present? Certainly these fictional accounts are fantastic examples told in dramatic ways to weave an interesting tale. But, what are the "realities" of technological control? Can it be, is it now, a controlling phenomenon in our lives? Is technological control as blatant or dramatic as the aforementioned stories make it?

Technical communicators have recently engaged some of the implications of technological determinism (although these discussions often have been couched in terminology other than "determinism," such as "power," or "empowerment"), and like the researchers in other disciplines, technical communicators also have a spectrum of beliefs on the topic of technological control. Many do not see a bright future because of what they view as the ultimate power technology holds over language and human action. Nancy Roundy Blyler (1994) makes this point quite directly in an article on technical communication's role in public empowerment, when she claims there is mounting evidence among technical communication researchers that, " . . . in fact, scientific and technological forces so manage discourse that full participation by citizens—and thus, to use Habermas' terms, emancipation and communicative action—are virtually precluded as possibilities" (Blyler 1994, 128). Despite this bleak outlook, however, Blyler concludes by calling for technical communicators " . . . to continue to explore the issue of empowerment and professional discourse, so that both the possibilities *and* [sic] the limitations of empowerment may be fully understood" (p. 142).

I agree with Blyler that the issues of technological determinism are not easily answered, and that they are issues technical communicators must engage. From the perspective of this book, such engagement is important in order to build a user-centered rhetoric that directly confronts questions of human control and technology. In a larger context, technological determinism is embedded in various assumptions that technical communicators have regarding technology. In the social arena, technical communicators (academic and nonacademic) deal with technology and people on a daily basis, thus making the practical issues of technological determinism an essential topic.

Plainly stated, I am claiming that we must pursue the phenomenon of technological determinism; it is ignorance or denial of technological determinism that we should avoid. While this may seem an absurd statement in some regards, it is nevertheless true that technological determinism is seen by some as mere fiction that can be erased by rigorously denying its existence. As I have said elsewhere, technical communicators and other language-related disciplines have a tendency to "romance" technology, especially when it falls into practical contexts.[5] In short, we should engage the debates of technological determinism in a forthright, yet deliberatory manner that will promote a critical, proactive awareness of technology's power. We should, in other terms, act as rhetoricians to explore, debate, and reinvent the issue of technological determinism in a human-centered way.

In the following sections, we will overview technological determinism as it is viewed by several disciplines that have engaged this issue for quite some time. The purpose of this disciplinary overview of perspectives on technological determinism will be to 1) illuminate respective methods and

---

5. See Robert R. Johnson "Romancing the Hypertext."

philosophies regarding technological determinism, 2) analyze and critique various notions of technological determinism, and 3) conclude with a discussion of what ramifications these differing views have for a user-centered rhetoric specifically and technical communicators generally. I would like to note that, like the analysis of human factors in chapter 4, this overview does not intend to be comprehensive. Instead, it will be selective and will involve a sampling of various research pertaining to technological determinism in these disciplines.

In the Eye of the Beholder: Sociology and the Denial of Technological Determinism

Social constructionism, also termed by some disciplines as simply *constructivism,* is based on the concept that reality is mutable, that there are no certain truths, and that knowledge is constructed through communally created knowledge and action. Technology to social constructionists, therefore, is a social construct that can be interpreted and reinterpreted depending upon the people involved, the context or situation in which it is designed, developed, or deployed, and the historical moment it resides within.

Although social constructionist theories have been around for about three decades (see Polyani; Geertz; Kuhn), the study of technology is a recent disciple of these sociological methodologies. The reasons for the slow turn to technology among social constructionists is open to speculation, but there are some interesting possibilities that reflect issues we have already surveyed regarding the history of technology and rhetoric. For instance, social constructionists involved themselves with the study of science well before they considered technology as a focus.[6] This parallels similar accounts by historians of technology, who claim that science held a dominant role as the discipline of choice among researchers.[7] Thus, there are a number of potential avenues for technical communicators to explore concerning the genesis of technology studies that could shed new light on the social nature of technology.

---

6. For an extended discussion of sociology's turn to science and technology, see Steve Woolgar's "The Turn to Technology in Social Studies of Science," *Science Technology and Human Values,* 16(1): 20–50.

7. See Layton 1971, *Mirror Image Twins*; also see Winner (*Black Box*) where, in reference to Woolgar (*Turn*) he refers to research that focuses on technology studies as "intellectual slumming" (p. 365).

As far as technological determinism is concerned, the social constructionists, most particularly sociologists, have approached it head on with a fervent desire to eradicate what they see as a simplistic and dangerous philosophy. As Wiebe Bijker, Thomas Hughes, and Trevor Pinch explain in the introduction to their collection of essays, *The Social Construction of Technological Systems,* one goal of the constructionist approach is to " . . . deny determinism" (12). Later in the same collection, Donald Mackenzie (1990) argues that his study of guided missile system accuracy is a clear example of socially constructed technology that discounts the " . . . simplistic technological determinism . . . "(195) of researchers who dismiss the social dimension of missile system development.

To argue against, and in some cases attempt to erase technological determinism, the social constructionists devote much time and energy to the pursuit of methodologies. In particular, they rely upon historical case studies and then use a variety of methodological tools to interpret the volumes of data they collect. The end of these studies, in terms of technological determinism, is often to demonstrate how certain technological artifacts have not been the result of efficiency or fixed historical sequences, but rather have been invented, designed, stabilized, disseminated, and eventually shaped by social forces. The findings they dredge and the stories they tell are often fascinating and rich with detail. Let us look at one significant avenue of social constructivist research through Trevor Pinch and Wiebe Bijker's analysis of the development of the bicycle to see how they conduct this research and what it can tell us about technological determinism.

The Bicycle: An Example of SCOT and "Interpretive Flexibility"

Bijker and Pinch use a rather formal method of analysis, which they term SCOT (Social Construction of Technological Systems). The overall goal of SCOT is to allow technical artifacts to be researched from a *multidirectional perspective* that places this mode of research on the extreme "soft" end of the determinism spectrum (see Figure 5.1). In addition, they argue that their method is a heuristic device derived from practical experience through case studies, as opposed to philosophical or theoretical foundations (p. 39). Their method is in part aimed at correcting some of the determinist tendencies previously mentioned in Staudenmaier's definition of technological determinism.

> In SCOT the developmental process of a technological artifact is described as an alternation of variation and selection. This results in a "multidirectional" [sic] model, in contrast with the linear models used explicitly in many innovation studies and implicitly in much history of technology. Such a multidirectional view is

> essential to any social constructivist account of technology. Of course, with historical hindsight, it is possible to collapse the multidirectional model on to a simple linear model; but this misses the thrust of our argument that the "successful" stages in the development are not the only possible ones. (Pinch and Bijker, p. 28)

The SCOT model described here moves counter to determinist views of technology by avoiding the linear sequence of history and by avoiding the success story that so often dominates the history of technology. Implied in this model too is an avoidance of concentrating on the lone great inventor. Instead of designating the invention as the moment when the genius inventor succeeds in stabilizing an artifact, Bijker and Pinch turn to the concept of *relevant social groups* that played a role in the development and dissemination of an artifact. These groups can be formal, like existing institutions or organizations, or they can be informal groups whose bond is the result of some common aspect of the artifact's existence.

In the case of the bicycle, they activate this method by first identifying large social groups, such as consumers or producers. Next, they narrow the definitions of the groups to provide a sharper focus for analysis. For instance, they examine women and men cyclists to see how they viewed the use of bicycles. They argue that women's clothing, and the social mores attached to wearing skirts and dresses, forced a bicycle design that was lower to the ground, thus keeping women's legs or undergarments from public view. For men, riding bicycles was seen as a "manly pursuit"—an attitude that for a time dissuaded the development of the inflated tire as these softer tires were not as dangerous, therefore not as "manly."

In addition to the actual users of the bicycle, Bijker and Pinch point to nonusers, or what they term in this case *anticyclists*. During this period (1880–1900) informal groups reacted violently to the new two-wheeled artifact. Citing examples of riders having sticks and stones thrown at them and being chased out of town by angry citizens, the two researchers make the case that the anticyclists found meaning in the bicycle artifact, albeit quite a different meaning than that of the users (p. 32). Such differences in meaning (as meaning is interpreted by the researchers) Bijker and Pinch define as *interpretive flexibility*, or the practice of using similar data but coming to quite different conclusions based upon various contextual constraints—a kind of relativistic tool, so to speak. For social constructionist advocates like Bijker and Pinch, interpretive flexibility is a mainstay of their approach to studying the development and impact of technological artifacts.

> . . . the interpretive flexibility of a technical artifact must be shown. By this we mean not only that there is a flexibility in how

> people think of or interpret artifacts but also there is flexibility in how artifacts are *designed* [sic].There is not just one possible way or one best way of designing an artifact. (*Social Construction*, p. 40)

The notion of technological determinism has little room in an approach like this. Set pathways of progress are constantly questioned, single interpretations of technological development are interrupted, and the potential for concentrating on consumers and users works counter to traditional ideas of the lone great inventor or genius developer. Thus Bijker and Pinch appear to promote what we might call a classic view of "soft" determinism.

There is a crucial problem with this conclusion, however, regarding the prospects for models like SCOT in a user-centered rhetoric. Bijker and Pinch clearly want to *deny* determinism, as they claim in the introduction to their collection. "With their emphasis on social shaping, Bijker and Pinch deny technological determinism. Borrowing and adapting from the sociology of knowledge, they argue that the social groups that constitute the social environment play a crucial role in defining and solving the problems that arise during the development of an artifact" (p. 12). To *deny* determinism, I take it, means to disallow its existence—to "write it out" of existence by demonstrating through a research methodology that it really does not exist. No room at the inn for technological determinism, they seem to be saying. Almost as a turn to a social determinism that eradicates its technological counterpart, SCOT appears to aim at the eradication of technological determinism rather than confronting it directly.

Such a position virtually puts SCOT off the "soft" end of the determinist spectrum, beyond the "softest" approach, as it attempts to make people the sole controlling agents of technology. Granted, this is a laudable goal and one that user-centered approaches would like to pursue. However, when one looks beyond artifacts as the sole technological component in the analysis, and instead includes institutions and large organizations as technologies, too, it is more difficult to determine what role "technology" is playing in the matrix. The SCOT model does not directly deal with these larger issues, at least not in any overt way. Thus, sociological approaches like SCOT are helpful to user-centered views to some degree, but they are inadequate when dealing with the politics or power of technology (at least as these methods have been practiced to the present time).

Bijker and Pinch attempt to argue away this deficiency by saying that SCOT suggests a way forward, a way to account for the larger issues of the " . . . sociopolitical milieu . . . [because by demonstrating] how different meanings can constitute different lines of development, SCOT's descriptive model seems to offer an operationalization of the relationship between the wider milieu and the actual content of the technology" (p.

46). But the evidence for this forward momentum is difficult to detect. The case of the bicycle, for example, begins to touch on these larger issues through the "anticyclists," but it ends with the claim that different meanings have been identified. Period.

Indeed, this shortcoming of SCOT may be a disciplinary issue. By that I mean that sociologists' methods do lend themselves to objective descriptions. Like historians in some respects, sociological researchers aim to tell *what*, but not very often *how*.[8] This is not to deny their research. It is essential and revealing. To use problem-solving methodology as an analogy, sociologists like Bijker and Pinch are invaluable in identifying the problem at hand. I guess it is up to others to develop a plan and move toward a solution.[9]

History and the "Turn to Culture" in Technology Studies

Historians are more content than sociologists to acknowledge the issue of technological determinism. This does not mean to imply that they are any more willing to accept determinism as a "good" force. Instead, it means they find technological determinism to be a reality that should be understood and identified, not just "written away." Possibly the most well-known scholar of determinist issues among historians is Lewis Mumford. Mumford, in an early volume of the journal *Technology and Culture*, described technology (or *technics* for Mumford) as existing in two parallel but unequal modes through the course of history.

> My thesis, to put it bluntly, is that from late neolithic times in the Near East, right down to our own day, two technologies have recurrently existed side by side: one authoritarian, one democratic, the

---

8. Bijker has continued to deal with issues of the larger political and social aspects of technology. In his second collection, edited with John Law, he concludes with a call for " . . . a tool kit—or rather a series of tool kits—for going beyond the immediate scapegoats and starting to grapple with and understand the characteristics of heterogeneous systems" (*Shaping Technology/Building Society*, p. 306). In a 1993 volume of *Science, Technology, and Human Values*, he states that, "My conclusion now is that the "social-constructivist program" can be pursued and extended in perspective, . . . in depth, . . . and with political relevance. . . . " (*Life After*, pp. 131–132). There remains, however, a static quality in his statements. "Tool kits" sound like more methods of analysis, and "can be pursued and extended" certainly implies that little has been done at present.

9. See Linda Flower (1993), Paul Anderson (1994), or G. Polya (1945) for a sampling of this approach to problem solving as it is used in composition studies, technical communication, and mathematics, respectively.

> first system-centered, immensely powerful, but inherently unstable, the other man-centered, relatively weak, but resourceful and durable. If I am right, we are now rapidly approaching a point at which, unless we radically alter our present course, our surviving democratic technics will be completely suppressed or supplanted, so that every residual autonomy will be wiped out, or will be permitted only as a playful device of government, like national balloting for already chosen leaders on totalitarian countries. (p. 2)

It is this characterization of technology of authoritarian and democratic—at once a devil and a savior—that surfaces in much writing on technological determinism by historians. Always weighing the balance, and often willing to admit to the power of technology as both an artifact and a sociopolitical force, the yin/yang of technology's power depicts the point of view of the technology historian. The skepticism of Mumford is also a defining element of the technology historian's mind-set. The warnings of technological power and the plea for intervention are quite often present in the stories of the technology historian. Rarely acquiescing to nihilistic bleakness, however, the technology historian usually finds hope in the power of the human spirit to live in a symbiotic relationship with technology.

Another characteristic of these historians is that they are more enamored with the technological artifact than are the sociologists, or the philosophers for that matter. Even though some sociologists are now calling for a return by their discipline to a greater acknowledgment of the artifact itself,[10] it has clearly been the domain of the historian to analyze and interpret the artifacts of technological development.[11] This love affair with artifacts, however, often has been criticized by those who argue that historians of technology should pay more attention to the cultural aspects of technology—that they operate from too much of an *internalist* perspective (see Staudenmaier, *Storytellers*). Just as historians of science have concentrated on great scientists and moments of discovery, historians of technology have a proclivity to investigate the lives of great inventors and their inventions. Such methods move from the inside out (hence the name *internalist*), from the artifact and its inventors out to the culture, and have been the focus of criticism by sociological researchers who advocate working from the outer

---

10. See *A Sociology of Monsters: Essays on Power, Technology, and Domination*, edited by John Law. In particular, see the introduction by Law and the chapters by Steve Woolgar and Susan Leigh Star.

11. For a classic work reflecting the focus on artifacts, see Charles Singer's five-volume collection written in the 1950s, titled simply *The History of Technology* (Oxford Press).

culture toward the internal workings of the artifact (called *externalist* methodology by the historians). The growth of the Society for the History of Technology and its journal *Technology and Culture,* however, has been a strong corrective during the past several decades to this criticism as issues of social influence are an increasing presence in the interpretations of technology by historians.[12]

To describe the technology historians approach to technological determinism, I have chosen a historical case study of farmers by Deborah Fitzgerald. This case, which of course is one argument, nevertheless depicts crucial and characteristic elements of a technology historian's disciplinary method and ideology as it pertains to technological determinism. It is, in short, a close analysis of primary textual sources, framed by an ideology that questions the power of technology in pointed yet positive ways.

Deskilled Farmers and the Silence of Technological Change

Technological determinism does not just have to occur through artifacts or organizational systems, but also can be made manifest through technique. This is a conclusion of Deborah Fitzgerald, a historian interested in farm labor and economics, who argues that farmers are the victims of *deskilling*: the loss of knowledge of techniques as new technologies or processes enter the work environment. Focusing on the shift from open pollinated seed corn techniques to the more "scientific" processes of hybridization, Fitzgerald argues that farmers were more profoundly deskilled as a result of hybridization than from any mechanized artifacts or systems (p. 327). A summary/scenario of her story of the deskilling of American corn farmers in the early twentieth century is presented next.

If you farmed corn in the U.S. during the early part of the twentieth century, you were most likely an expert in the art of selecting seed corn for the following year's crop—known as the technique of open-pollinated seed production. The knowledge needed to select the most appropriate seed included maturity rates, climatic conditions, soil quality, bug and disease prevention potentials, and the value of the crop on the open market. In addition,

---

12. Although I have referenced it several times already, it would behoove anyone interested in the development of the history of technology to look at John Staudenmaier's *Technology's Storytellers.* The book thoroughly chronicles the development of the field through the lens of *Technology and Culture,* the international mouthpiece of the discipline.

this art required you to keep consistent and accurate records, or "score-cards" as they were sometimes called, so that you would know from year to year, and even decade to decade, what your seed selection had accomplished in terms of production, disease prevention, and so on. The knowledge of these arts of open pollination was so pervasive that 99.4 percent of the corn grown in the U.S. in 1933 was produced from seeds selected through this "farmer-owned" knowledge.

By 1945, however, this art of seed selection had been almost totally supplanted by hybridization—a technique of seed production that is carried out by agricultural scientists under the auspices of large corporations or university experimental farms. In fact, hybridization became so popular that by 1945 hybrid corn constituted 90 percent of the corn grown in the U.S. Thus, quietly, quickly a type of knowledge, specifically a type of user knowledge, was gone—transferred and transformed in historically the "blink of an eye."

Fitzgerald's research, although it deals with more than just determinist issues, is an excellent example of how technological determinism is manifest on several levels. We have already had a glimpse of how this shift of techniques affected users (the farmers) directly, thus demonstrating how technology or technique can determine the knowledge of a social group. In addition, her study examines the role of other social groups—the seed companies, the government, and agricultural research universities. A most revealing point is made in reference to the power struggle that went on between the corporate seed producers on the one hand, and the farmers on the other. In the early days of hybrid development, some scientists believed that the farmers would be the best source to judge high-quality corn strains because they had, for generations, had so much experience with seed production. At the University of Illinois, farmers were subcontracted to do breeding work, partly because the land resources of the university were not big enough to do all of the development and experimentation "in-house." As Fitzgerald points out, the use of farmers as hybrid seed producers was not widely applauded by the commercial breeders.

> This did not sit well with the commercial breeders, who felt that farmers could not be relied on to keep accurate records or to keep the lines pure. At issue was the question of whether ordinary farmers were competent to manage the crossing of corn. Twenty years earlier this was a task that Wallace[13] had claimed "anyone" [sic] could do. . . . [Nevertheless], [c]laiming that farmers were

---

13. The owner and editor of *Wallaces' Farmer*, a popular journal of agricultural news and techniques during the early twentieth century.

> not really capable of producing good seed themselves, commercial breeders tried to think of ways to stop such practices. . . . In essays titled "Beating the Farmer at His Own Game" and "Discourage Home Seed Saving," seed producers expressed their growing frustration. (*Deskilled*, 336–38)

The commercial seed producers triumphed as open pollinates were systematically driven from the market. They even furthered their claim to the ultimate knowledge of corn seed production by suggesting that the choice of which hybrid strain to choose (hybrid strains can be "temperamental" from locale to locale) should be left up to the commercial producers. As one seed catalog stated, "You may not know which strain to order. Just order Funk's Hybrid Corn. We will supply the hybrid best adapted to your locality" (p. 339).

Analogies to user knowledge as discussed in a previous chapter are evident in the disdain for farmers' knowledge by the commercial interests. Not only was their knowledge undervalued, it was literally usurped by the hybrid experts. The farmers, once the experts in seed production, were now relegated to novice status in less than a generation. Ironically, the knowledge the farmers once owned they now were forced to purchase, and at prices I would assume were set by the commercial producers.

This example of technology's power is also interesting as a contrast to the "anticyclists" of Bijker and Pinch. The anticyclists were disruptive in their reaction to the new technology of the bicycle. Seeing it as a threat to their social sphere, they openly attacked the technology and its users in a true Luddist fashion. The farmers, however, lost their knowledge without much resistance. The reasons for this lack of resistance are complex and can be attributed to several factors. One factor is the probable "acquiescence" of farmers to the less labor-intensive process of hybridization—they no longer were bound to long hours of repetitive tasks and record keeping (p. 326). Another factor could have been the growing faith by twentieth-century Americans in the promise of science as a way to improve life. Nevertheless, farmers did lose knowledge and possibly much more. Not unlike indigenous cultures that lose their knowledge, their land, and ultimately their fundamental culture to more powerful forces of domination, social groups like farmers can lose a way of life to technological change.

Like the sociologists, however, the historians seem less than willing to offer pointed critiques of particular technological issues for the purpose of advocating change to our current political or social structures. There are, of course, exceptions. For instance, Lynn White included a chapter[14] on technology assessment in the United States and argued that historians,

---

14. See chapter 16 of *Medieval Religion and Technology*.

because of their knowledge of past problems with the dissemination of technology into cultures, could lend a helping hand to future government assessment efforts of technological impact. There are also grand gestures toward changing society through historical study, such as Mumford's plea for democratic technologies: a worthwhile plea and one that deserves the most serious attention, but one that is difficult to operationalize in a meaningful way. For the most part, the concept of proactive scholarship that offers specific suggestions for intervention appears to be absent from the technology historian's method, and instead the aim of just describing the past is the primary goal.

Beholden to Technology: Philosophy and the Political Context

The philosophy of technology, similar in some ways to technical communication, is a discipline looking for an identity. This is ironic considering that philosophy as a discipline in general has a longevity at least as long as history's and certainly longer than sociology's. Nevertheless, the philosophy of technology as a discipline has been somewhat slow to materialize. I would take issue with Bijker and Pinch, though, (who cite R. Johnston as evidence) when they claim that the "literature on the philosophy of technology is rather disappointing. We prefer to suspend judgment on it until philosophers propose more realistic models of science and technology" (*Social Construction*, 1987, p. 19). It is not the *literature* by philosophers of technology that is problematic. With philosophical research of technology conducted by the likes of Jacques Ellul, Jose Ortega y Gasset, and Martin Heidegger through the twentieth century, and most recently by Donald Ihde, Langdon Winner, and Carl Mitcham (to name only a few), it is difficult to claim that the literature is disappointing. Instead, it appears that the philosophers of technology have had a problem negotiating a common ground of research upon which to clearly base a disciplinary research agenda in the same way historians or sociologists have. Carl Mitcham suggests that the philosophy of technology has had difficulties forming because it consists of two parts that he likens to ". . . fraternal twins exhibiting sibling rivalry even in the womb" (*Thinking*, p. 17). The twins he refers to are the philosophy of technology as an engineering discipline and as a humanities discipline—one pro-technology, the other more critical.

Clearly, we will be looking at the humanities side of the divide in this present discussion: a side that deals with the political implications of technology. While not all philosophers attend to the political aspects of technology, it is evident to me that they do so more seriously than either the historians or sociologists. The philosophers have their weaknesses, too, and therefore my intent is not to create a hierarchy that places one discipline in a more acceptable light than another. It is important that a user-centered rhetoric of technology draws as widely and carefully as possible in its disciplinary journeys. Therefore, the aim of this final overview of disciplinary views of technological determinism will be to present the philosopher's arguments of technology and politics so we can wed these political dimensions with the more descriptive dimensions of the two previous disciplines of sociology and history, thus moving toward a preliminary pluralistic synthesis of views that will be useful as we apply the user-centered concepts in the final chapters.

On the "hard"/"soft" spectrum presented earlier (Figure 5.1), philosophy, probably more than the other disciplines we have discussed, has adherents clearly on the "hard" side. Ellul is without a doubt the most famous of the near-nihilistic critics of technology. For Ellul, it is not modern technology that is the determinist culprit per se, but it is the ancient concept of *technique* that he claims has been so pervasive throughout the history of Western society and has been turned uncontrollably against the spirit of humankind. Ancient societies, according to Ellul, allowed individuals to have a choice regarding techniques, and in those times technique was something you could live with. Ancients could choose which techniques they wanted to use and which they preferred to ignore.

Modern society, however, is unquestionably different for Ellul. Now, technique has become a corrupting force that turns existence into a technological order that controls the destiny of humankind through the creation of a common view of life—a view that is "in the eye of technology."

> It [technique] has been extended to all spheres and encompasses every activity, including human activities. It has led to a multiplication of means without limit. . . . Technique has been extended geographically so that it covers the whole earth. . . . Moreover, technique has become objective and is transmitted like a physical thing; it leads thereby to a certain unity of civilization, regardless of the environment or the country in which it operates. (*Technological Society*, p. 78)

Technology, in Ellul's estimation, is a living thing that can (does) control every aspect of human existence. The only hope for humankind is to regroup and start again. Ellul's solution, in brief, is to reconstruct a world with a forgiving God, where technique will once again be under the control of humans.

There are few philosophers who adhere to the strong determinist stance of Ellul. But many do hold to the assumption that technology does have determinist strands, and powerful ones at that. Some of the most powerful, it is argued, reside in the political sphere where technology operates on many levels and to many ends. In this final discussion of disciplinary views of determinism, we will look at a piece by Langdon Winner that asks the question, "Do Artifacts Have Politics?"

The Politics of Technological Determinism

Winner's work has been a central force in the philosophy of technology since the publication of his book *Autonomous Technology: Technics Out-of-Control as a Theme in Political Thought* in 1977. In this book he surveys the history and philosophy of technology for the purpose of outlining the issues that arise around philosophy, technology, and political action. At the core of this book are questions about the "Ideological presuppositions in radical, conservative, and liberal thought [which] have tended to prevent discussion of a problem in technics and politics that has forced itself onto the agenda of our times" (*Autonomous*, p. x). The "problem" Winner refers to is *limits*: the points at which we stop the gears of technology and reflect upon what it means to live in a technological society, what it means to admit that technology is a viable, political force that we must live with, instead of live for.

In his second book, *The Whale and the Reactor*, Winner continues to explore the theme of limits and politics through a series of closely knit essays. In "Do Artifacts Have Politics?" he analyzes ways that "artifacts can contain political properties" (p. 22). I will concentrate on two examples he uses in the following scenario:

Robert Moses, the legendary director of New York City's public works during the early twentieth century, was famous for his vision and sense of purpose. Nowhere is this more apparent than in the many bridges and overpasses he engineered that link Manhattan to the boroughs. When you look at these thoroughfares, however, you are struck by the low overhead clearance of these structures—only passenger cars and relatively small trucks can pass under them.

You might be inclined to think that this was an engineering quirk, or that there was some constraint, or maybe that it was just an oversight made before the days of larger vehicles. In essence, you would be right on one account—there was a constraint, only it was not a technological constraint that any engineering techniques could overcome. Instead, the constraint was Moses himself. Robert A. Caro, Moses' biographer, demonstrates that these low overpasses result from Moses' deep class and racial biases. He wanted to limit the flow of passenger buses into Manhattan—for the purpose of limiting access to Manhattan of those who rode such public transport, namely minorities and the working classes. He even went as far as stopping the building of a railroad to Jones Beach, a public park that Moses prized as one of his finest accomplishments: one he wanted to protect for the middle and upper classes.

The mechanical tomato harvester is a machine that has been a boon to the tomato industry, at least from the perspective of those in the political power seats. Developed in the late 1940s as a joint effort of the tomato industry and the University of California, the picker saved the industry about five to seven dollars a ton as compared to hand picking techniques (in early 1970s' dollars). The machines cost approximately $50,000 apiece and were thus affordable by only large tomato growing operations.

The effects of the mechanical harvester on tomato growers and labor workers was, as you might expect, profound. Because of their high cost, they eventually forced many smaller growers out of the business, reducing the number of tomato growers from about 4,000 to about 600 in a thirteen-year period (1960–1973). The number of labor jobs fell as well, dropping by approximately 32,000 by 1980. A lawsuit was filed against the University of California by an organization representing the workers, claiming that tax monies were being used to the detriment of workers, farmers, and others. The university countered the charges, claiming that they would have to stop all research that aimed at "any practical application" if they were to acquiesce to the charges.[15]

The politics of technology—by that I mean the power relationship between users, producers, designers, and the artifact—is evident in both of the aforementioned cases, but it presents itself in two distinct forms. In the New York freeway example the political power appears to be conscious. Moses made decisions based on overt ideological assumptions of who should benefit from his oversight of technological development. By denying access to the public places he controlled, Moses effectively used technological design, and then the resulting artifacts, to dictate to a great degree what the "look" of New York would be. The politics of technology is clear, blatant, and undeniable, even though it might not be evident in the physical artifact when casually observed.

---

15. For more on the issue of land grant colleges and agricultural research, see Wendell Berry's *The Unsettling of America* and James Hightower's *Hard Tomatoes, Hard Times.*

The tomato harvester, however, is somewhat different as it depicts technological power as more covert. Here appears Staudenmaier's definition of technological determinism, as the desire for efficiency and progress are paramount in the design, development, and diffusion of the artifact. The politics of the technology in this case is embedded in the machine, and although it may not be as blatant as Moses' politics, it nevertheless has a similar consequence. The technology is determining the decision making of the *polis* to a great degree, despite the fact that the university, in a back-handed kind of way, admitted its guilt in the matter through the implication that they would have to discontinue all "practical research" if it led to questionable social outcomes. After all, the university argues, how can we stand in the way of progress? It is inevitable. Echoing strains from Staudenmaier's definition of technological determinism, the university invoked arguments of the "necessity for forward progress" in the service of "the technological success story," essentially setting up a binary between action and inaction as the only possible alternatives, with action in the service of technological progress, of course, holding the winning hand.

These historical examples are only a few that can be drawn to demonstrate the political power of technology. We could dismiss such power as inevitable, or maybe even inconsequential if we want to believe that historical reflection of this sort is merely a form of "nay-saying" or possibly inaction that ignores the concerns of such issues as economic development. But before we deny the existence, or before we let the "powers that be" just run the technological show, consider one more example—the case of the "information superhighway." In an article titled, "Technology and Democracy," Vice President Al Gore outlines his vision for a National Information Superstructure (NIS), " . . . a seamless web of communication networks, computers, databases, and consumer electronics that will put vast amounts of information at users' fingertips and will forever change the way we live, learn, work, and communicate with each other" (p. 42). Gore continues by saying that the NIS should be expanded to an international initiative that he dubs the Global Information Infrastructure (GII). The GII, then, would

> . . . be a metaphor for democracy itself. Representative democracy does not work with an all-powerful central government, arrogating decisions to itself. That is why communism collapsed and apartheid fell. Instead, representative government relies on the assumption that the best way for a nation to make its political decisions is for each citizen to have the power to control his or her own life. (p. 42)

So we return to the issue of the power of the individual person, to the problem of the one and the many, as Carolyn Miller puts it.[16] We also visit the landscape that Jacques Ellul fears to tread upon, but ground that we nevertheless seem to be committed to traversing—a land where control is ambiguous, but where we delude ourselves into feeling that we, not technologies, are the controllers. When I read Gore's vision, I find myself working as a traditional rhetorician: as someone who looks at language and sees the power that is inherent in it. I see language that cries out, "It's not *we* who are controlling this change, it is the technological artifacts and the systems they comprise." When Gore says that the GII will "promote the functioning of democracy," "promote economic growth," "revolutionize the world economy," "promote the ability for nations to cooperate with one another" (p. 42), I want to agree, I want to believe that the technology is in our control—in the hands of the citizens. Yet it is not that easy or clear-cut.

Here we are getting to the issues central to a user-centered rhetoric, but I fear that I may be straying too far off base here, especially in that I am dabbling in the realm of contemporary politics. However, my point is to expose the nature of technological determinism in our everyday experience—as it might pertain to our working lives, or in this case, as it pertains to the future of federal or international government policy. My criticism of Gore is not that he should immediately abandon his vision of an international communications network. Rather, I have presented his "technology philosophy" as a way to ask, 1) whether his vision is illusory: a figment created by a naive allegiance to technology in service of the common good or democracy, and 2) if it is illusory, then can the illusion be refigured to represent more of the human element and less of the technologically determined one? In other words, can *we* determine technology? This, I believe, is the focal question that a user-centered rhetoric should address in regard to technological power.

Language and power, people and technology, the *polis* and the individual—they are inextricably intertwined in a matrix. But that matrix should be more negotiated, more flexible as far as all participants are concerned. We need to, in other words, be able to activate the negotiation so that citizens of the polis, the end users of technology, can truly have more impact upon the technological systems.

A dream? Idealistic? Possibly. But there are some avenues to pursue in an attempt to promote such change, and they are in part found in a refiguring of what we mean by users and use. For instance, a pluralistic approach

---

16. See Miller's "Rhetoric and Community: The Problem of the One and the Many."

to technological study, one that borrows reflectively from the aforementioned disciplines (and other disciplines too), can provide insights into technology that are deep and revealing. Are there, for instance, "anti-computerists" we should analyze to understand their ways of manifesting their existence? Do these anti-computerists smash machines, as some have done recently in open public displays,[17] or are they more quiet and subdued as they merely shut off their machines and walk away in frustration? How will these end users get on the information superhighway, how will we know where they exist? Are they like the corn farmers—silently deskilled and eventually disrupted from a particular way of life?

In addition to developing pluralistic research methods, we also can act locally by applying such methods in our everyday experiences. Through this last scenario of a trip I took through a new high-tech communication classroom in a business school, I will consider how technical communication teachers might rethink their role in rethinking technology.

The "state of the art" facility was elegant. Supported by a private donor who wanted to ensure that the future graduates of the school would have experience with the newest computer technologies, the hallways and rooms looked more like a corporate office complex than a university classroom area. The largest room of the facility, where most of the instructional activity took place, consisted of three horseshoe-shaped rows, each one raised slightly higher than the other, resembling something of a theater-in-the-round. The horseshoe arrangement faced a large cinematic screen and a gadgetry-bedecked podium. Each row contained about ten padded, theater-like chairs, and in front of each chair was a computer terminal, half submerged in the horseshoe-shaped desk. The tour guide motioned for us all to take a seat. She then proceeded to strap a small device to her wrist and, pressing a button on the device, "commanded" the electric window shades to descend. The room was suddenly in complete darkness, save for the glow of the computer screens at each seat.

"The main purpose of the room," said the tour guide, "is to bring the business students into the twenty-first century by providing them with the technology they will encounter once they are 'out there'." Pointing to the large projection screen at the front of the room, she once again activated, with a flick of the wrist, an overhead projector that immediately began showing an economics lesson. We sat and watched as the screen displayed

---

17. See Steven Levy, "The Luddites Are Back."

a multimedia lesson that incorporated full motion video, animation, color, and even the instructor's voice as he explained various economics concepts.

When the lesson was over, the guide opened the session to questions. Several participants asked questions about the technology—What was the software? How hard was it to learn? What support was there for learning about it? A few asked about the students' response to the technology—Did they like it? Did they have problems learning how to use it? One member, however, asked a somewhat different question: "I understand that many modern businesses believe that the most important thing for students to know is how to collaborate, how to work with others in teams. How will a room like this one enable students to get such experience?" "Oh, quite easily," replied the guide. "They will be able to communicate with each other, the teacher, and, in fact, with anyone in the world through the computers at each seat. They will, I suspect, collaborate more than they ever have before."

This scenario of my trip to the high-tech computer classroom exemplifies some of the complications that user-centered theory must confront regarding technological determinism, in particular, what teachers must face as computer classrooms become the commonplace environment of learning. Is an electronic classroom such as the one previously described designed to promote learning? If the goal is learning rote facts presented by an expert (the professor) to learners (the students), then I am quite sure it will fulfill the purpose. If, on the other hand, the learning goals are based more on concepts of collaboratively constructed knowledge, then the answer is more uncertain. Will the collaboration that takes place in this classroom really be collaboration? In other words, does the technology change the nature of collaboration, or does collaboration stay pretty much the same as before the computer? More fundamentally, will collaboration through the computer actually instruct the students in the arts of collaboration, or will the new technology merely provide the appearance of learning about collaboration? What is the end of the collaboration—to learn, communicate, or use the technology? I ask questions about the historical moment that we are in. Are we reflecting upon the impact of the technology in our current context? Are we, as we refigure collaboration, for instance, designing the technology to aid the collaboration, or are we allowing the technology to refigure collaboration for us? We might also ask questions of the rhetoric that is used in these situations. What does it mean "to move into the twenty-first century?" What does it mean, in fact, "to move into?" Does such inevitable forward progress happen out of necessity, or can it be controlled, altered by the users who will ultimately interface with the technology?

These are questions with no clear-cut answers. Yet, it would be difficult to develop a user-centered theory without confronting technological determinism. The term *user-centered* implies that systems should be designed for human users, and it further implies that users should be participants in the design and development processes. However, in much user-centered design, the issue of determinism is not dealt with directly, or with any sense of historical context. To be blunt, for the most part, technological determinism is ignored. The reasons for not confronting the issue directly might be due to system complexity, or to practical constraints like getting a product "out the door." Nevertheless, any attempts to place users centrally into technological design, dissemination, and implementation must confront the issue of technological determinism, even though the result can be a problematizing of the theoretical goals. We can enact some change in our everyday life, whether in teaching, research, or the workplace, but enactment begins with a reflective sense of the problem at hand. Unreflective assumptions about the possibility of human control over technology are problematic, if not downright dangerous. As technologies have become more complex, more pervasive, more of "what we live," they have sparked debate and concern over the controlling nature—the autonomy—of technology in the modern world.

Technology is alluring. It can be beautiful, helpful, and even lifesaving. Literally or figuratively, it can take us places and give us things. Computers get us on the "information superhighway," cars get us on the "real" superhighway; the technology of language can persuade others to give us what we want; language can help us transcend mundane, everyday experiences. Technology can even transport us into the afterlife, either by ancient chariot or by a more modern form of transportation. For the Mississippi Delta blues player Robert Johnson, 1930s' technology represented a final state of grace where one can endlessly tour highways and escape the harsh realities of cotton fields, lynchings, and jealous lovers.

> You may bury my body
>   down by the roadside.
> You may bury my body, ooh
>   down by the roadside.
> So my old evil spirit
>   can catch a Greyhound bus and ride.
>   *Me and the Devil Blues,* recorded in 1935

In this way technology is a tool—there to fulfill our needs, at our command. Hence, when technology is a tool we are controlling, it is the power that we want when we need it. Technology, however, can be controlling. There is the powerfulness that resides within the tools and systems we either use or are a part of. Technology is, like rhetoric and fire, a paradox of power and powerfulness. Technology may be the defining paradox of our culture: a paradox that, like virtually all paradoxes, we can neither escape nor ignore.

PART III

Communicating Technology

CHAPTER 6

When All Else Fails, Use the Instructions

Local Knowledge, Negotiation, and the Construction of User-Centered Computer Documentation

> Consider the "stop print" function in some database programs. You make a mistake or you see that it's not printing the way you want and you want to kill it. So you quickly open the manual under "print" and look for "stop" and there is nothing. So you look under "k" for "kill"; you even look under "a" for "annihilate." And there is nothing in the index that tells you how to make the darned thing stop.
>
> There should be a law: No one who hasn't managed a database should be allowed to program one. I call this law: "No one should be allowed to make menus who hasn't had to eat off them."
>
> —Ellen Bravo, *9 to 5*

Users of technologies, at one time or another, use some form of written instructions to help guide them through the "correct" procedure of whatever artifact they are currently using. Whether we are putting together a barbecue, repairing a bicycle, tuning a car, or following the directions for taking a medical prescription, we often are compelled to consult the instructions: the texts[1] that contain the knowledge of what it is that we want to do. Instructions come in many shapes and sizes, but they need not be complex or voluminous to cause confusion or frustration. Early studies of how people comprehend warning signs (see Chapanis 1965; Hartley 1985) indicated that even the most rudimentary information can be misleading or ambiguous. More recent investigations have demonstrated that the brief but crucial information included on drug prescriptions often is not followed correctly, thus leading to possible medical complications and potential lawsuits (Williams, et. al. 1995). Clearly, despite their ubiquity, written instructions have something less than a glowing reputation.

Nevertheless, as much as we may try to avoid them, we simply cannot escape the presence of instructional texts in our everyday lives. In some instances, such as those involving computer technology, we might not actively seek instructional assistance but it is offered to us anyway. For example, I unexpectedly received the following message while loading a new piece of software on my personal computer:

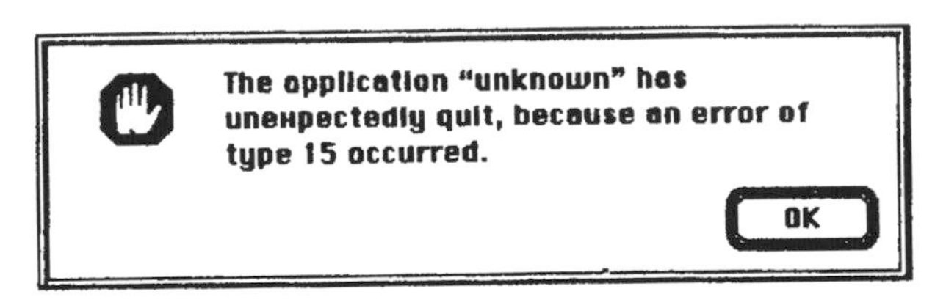

Unless I am a super user of the system, or I happen to have an up-to-date definitive reference document explaining all of the error messages of the operating system, there is little useful information contained in this computer displayed "dialogue box."[2] It provides no visible knowledge of why the system stopped working or what the error might be. Was it my error or the system's error? Am I compelled to agree with the machine and just commit myself to agreeing that everything is "OK?" As the authoritative upraised hand might suggest, am I to fear that by choosing "OK" I might be

---

1. In this chapter's discussion of instructional documents, "texts" will be used as broadly as possible: printed pages, on-line screen information, medicine bottle labels, road signs, explanatory cartoons—anywhere instructional material might reside.

2. The term *dialogue box* is curious in that it actually represents a monologue presented through the computer. Probably this terminology occurred because the computer has been characterized as an intelligent machine that communicates with people. At the same time, it also appears to be a reflection of a false sense of the reality of human–computer interaction, discussed in chapter 4.

harming the machine or the software? Above all else, if I attempt the procedure again (which I did), and I receive the same message again (which I did), what am I supposed to do?

Instructional materials have more far-reaching consequences than merely frustrating or confounding a user. James Paradis's study of the litigation that surrounded the operating manual for a direct-acting stud gun is but one example of the legal ramifications of written procedures (1991). Such discussions of the legal issues associated with instructions have given me pause as I have gleaned numerous documentation over the course of my research. Possibly the most frightening example was the procedure for cleaning a rifle. In the fourth step of the procedure a warning (all in upper case letters) stated:

4. NEVER WORK ON A LOADED GUN!

It appears that the authors of these instructions assumed the user would make it to step 4—a deadly assumption, given the context.

Curiously, with the vast number of texts dedicated to instructing humans *how to do*, the potential legal problems surrounding them, and the significant effort that several disciplines have devoted to the study of instructional materials, it is ironic that the reputation of these texts remains so dismal. Try to remember the last time you heard someone say, "Wow! Those were really fantastic instructions!" Chances are you can not recall such a statement, and if you can then it was probably pure sarcasm. Why have instructions received such a "bum rap?" Why are they used only when "all else fails?" Why do instructions seemingly fail to do what they are supposed to do—instruct humans on how to use technologies?

Part of the difficulty with instructional materials is due to their deceptive simplicity. Instructions are meant to make the assembly or use of an artifact appear simple: they are a masking device for the complexity of systems and artifacts. Instructions are expected, as the renaissance courtier would have said, to possess *spretzaturra*—the ability to make the difficult appear mundane. This aim of instructions is primarily persuasive, as it can promote the commercial marketing of the technological products they accompany because it makes them appear easy to use. Ironically, the apparent ease of use that the texts display makes the instructional documents themselves look "easy." They look simple, so they must be simple to write (and simple to use, thus invoking the image of "idiot-proofing," discussed in chapter 3).

Further, instructional documents are what visibly reside between the user and the black box of technology. Therefore, they play the role of messenger in the transfer of the technology to the user and are more apt to be blamed for any breakdowns that occur during use because they are present at the moment of frustration or breakdown. As such, instructions are intruders into users' spaces. Humans have a proclivity to approach technologies without instructions in hand (literally). We have a strong desire to "get going" when we are using new artifacts, and instructional texts only intrude on our mission. Technologies are *means* to our ends, after all, and the activity of using instructions can be an additional barrier to achieving our goal. That is, when we use a technology we are too busy trying to figure out how the artifact can help us obtain our goal, and the instructional document that accompanies the artifact is perceived as a time-consuming nuisance.

Perhaps, as commonly suggested, the poor reputation of instructions can simply be blamed on bad writing. Many technical communication textbooks proclaim that the lack of clarity, brevity, or syntactic quality can be a major downfall of effective instructions. The evidence is too strong, however, to blame the monolith of poor writing as the essential problem of unusable instructions. As I mentioned at the opening of this chapter, readers can often comprehend the syntactic structure or the vocabulary used in instructional materials, but they cannot complete the intended actions that the instructions are meant to support.

Indeed, the aforementioned problems contribute to the difficulties we have with instructional documents. However, as accurate as these complaints may be, they do not actively address the fundamental *end* of usable documentation: the use of technology by users.[3] Instructions have been written with the artifact or system (or the marketing of the artifact or system) as the focus, and the *end* of written instructions has thus been to merely describe the "thingness" or sleekness of artifacts and systems instead of their situated use. *There is, in short, a deeply embedded assumption that instructional materials are adequate merely because the information is there in either print or on-line form.* Never mind where or how the instructions will be *used*, this assumption dictates; the fact that users have a text in front of them is enough. Ironically, almost insidiously, this assumption places virtually the entire burden of comprehending instructional text on the user.

For instance, in a recent study of medical patients, it was concluded that there is something akin to a literacy "epidemic" in regard to the information with which people are provided as they seek medical care. When patients at two large urban hospitals were studied concerning their ability to comprehend written materials provided by medical personnel, it was concluded that many patients suffered from "inadequate functional health

3. Here I am referring to the "end" of technology as being in the user—see chapter 2 for an elaboration on this concept.

literacy" (Williams, et. al., p. 1677). As the researchers explain, "Functional literacy is the ability to use reading, writing, and computational skills at a level adequate to meet the need of everyday situations. Functional literacy varies by context and setting; the skills of a patient may be adequate at home or work, but marginal or inadequate in the health care setting" (pp. 1677–78). In the study, they found that patients had difficulty comprehending, remembering, or acting upon medical information contained on such things as appointment slips, birth control instructions, prescription dosage information, and consent forms. With the exception of the consent forms, which the researchers admit are difficult to read due to legal concerns regarding malpractice (see p. 1681), the study contends that most of the documents used in the study are simple texts that are written at or below a sixth-grade comprehension level. In other words, the assumption here is that regardless of context (even though these authors seem to admit the problem of context in their aforementioned definition of functional literacy) the "medically illiterate" are just plain unable to read and write. I guess the same can be said of the "computer illiterate," who cannot understand simple texts like the error message shown on page 116. The information from within the medical black box is made visible, but those who cannot comprehend it are illiterate. Heaven forbid that the system should be changed to expose the black box! The knowledge of the system is assumed to be correct in the present arrangement. In addition, the users have no business taking part in the decision-making processes of the system, because they do not understand the "complexities" of it in the first place.

"User beware!" is the appropriate slogan. The user is relegated to the position of a one-way receiver who has little knowledge of the technology itself or how the technological system might be refigured through an active negotiation of designers, producers, and users. Instead, the situated activities associated with use are supplanted in favor of the static, *correct* description of technology, *ala* the knowledge of the "expert" who designed and developed the artifact. Thus instructional materials have, innocently or not, played a significant role in the continuation of the modern technology myth that the role of experts is to invent, while the role of novices is to await, with baited breath, the perfectly designed artifact. The complicity of instructional documents with this top-down, expert-to-novice ideology has a number of crucial ramifications that have been mentioned in previous chapters regarding the determinist tendencies of technology development, the perception of users as "idiots," and the use of the word "human" in technology development to give credence to technology-driven initiatives. Most obviously, a disproportionate binary is created that at once empowers the developers and disseminators of technology while it disempowers the users who receive the technologies at the end of the production cycle. In this chapter, I will argue that instructions are more often than not an afterthought of the technology development process. While this in itself is not a revolutionary claim, the consequences of this arrangement are substantial.

I have chosen computer documentation as an example of instructional text development for several reasons. First, the problems associated with instructional text have been magnified as a result of the personal computer. The computer has emerged into our culture through numerous contexts, many of which include people who have had no previous experience with computer technology. Even though these users may be highly educated, they received little, if any, formal instruction concerning computers in their traditional educational experience. Consequently, computer documentation has become the primary vehicle for educating this wide variety of users about this rapidly developing technology.

Second, computer documentation resides in more than one medium (print and on-line forms), and thus further complicates the challenge of user-centered theory. Such multiple media involvement opens doors to a number of fascinating and fundamental questions of reading, writing, text design, human comprehension, and the concept of self, to name only a few. In particular, the differences in media illuminate problems with traditional instructional genres. As we shall see later, traditional genres often do not address the needs of users in the ever-changing contexts of computer-mediated media.

Third, computer documentation writing is arguably the largest source of employment presently for technical communicators. This in itself offers more than one reason to seriously pursue computer documentation as it brings to the front questions regarding the practice of writing, the role of writers in industry, and even the position of power that technical communicators hold in institutional settings. For instance, technical and scientific writers have traditionally been defined by the discipline they work within—such as medicine, environmental science, industrial technology, and so on. The computer, however, complicates these divisions as many writers now perform functions that were once primarily within the jurisdiction of the computer documentation specialist. Most technical and scientific writers, regardless of their specialty, write computer-related instructional materials in print and on-line forms because their audiences are increasingly using the computer medium as a text.

Fourth, computer documentation is a marginalized text in the sphere of academic research. By that, I mean it has seldom received the serious theoretical, historical, or general scholarly attention that many other forms of texts have received.[4] Here I am not just including literary texts (like novels, poems, short stories, diaries, essays, etc.) as the favored sites of analysis, but I also include scientific articles, laboratory reports, legal documents, and grant proposals. These texts have been given rigorous attention at one time or another by scholars in a variety of disciplines.[5] Computer documentation is one of the most complicated forms of text we possess—both in terms of product and process. It simply deserves a serious analysis apart from the usual "applied" discussions that usually revolve around it.[6]

Finally, computer user documentation is a valuable lens, not only for the study of the texts themselves but also for studying the users who use them and the constituent cultures[7] that arise/evolve from the activities associated with computer technology. It is, in fact, the social arrangements and processes of computer documentation development that will provide much of the focus for the upcoming discussion.

In this section, three approaches to computer documentation development will be discussed: system-centered, user-friendly, and user-centered. These three terms have been borrowed from various places, but the use of them as a taxonomic device to describe computer documentation development processes is my own invention. Therefore, a definition of each term as it applies to the current discussion of computer documentation theory will be provided here.

---

4. The most notable exceptions are studies like Stephen Doheny–Farina's or Roger Grice's about documentation processes in industry settings.

5. A short list of work done with these other forms of "nonliterary" texts would include Greg Myers' *Writing Biology*, Bruno Latour and Steve Woolgar's *Laboratory Life*, Bazerman's *Shaping Written Knowledge*, Yates' *Control Through Communication*, Dorothy Winsors' (and others) work on the Challenger disaster.

6. Apart from the practical discussions of writing computer documentation, most research using computer documentation as a textual "site" has employed it in the service of explaining visual presentation of text (see Kalmbach or Kostelnick), or as a model of products that exemplify usability methods (see Dumas and Redish).

7. "Culture" is used broadly here to mean any community that might have common bonds due to context or practice: that is, workplace cultures, classroom cultures, and even electronic cultures such as those emerging from the use of the Internet or other computer networked arrangements.

Artifacts, Experts, and Idiots: The System-Centered View of Computer Documentation

This view of computer system development has dominated the field of computer science for many years, and it has driven much of the technological development that computer systems have enjoyed in recent years. The reality that computer systems and applications need user documentation, however, has sprung from the recent phenomenon of the microcomputer and the proliferation of end users. System-centered design practices, in contrast, have evolved from an earlier time in computer technology where users were experts in the use of the systems, thus had little need for instructional texts in computer usage.

Documentation that was developed from these system-centered practices was aimed more at the static description of the computer's components and features than it was at any kind of end use of the computer. A now classic example of this is the huge library of documentation that accompanies UNIX, an operating system for mainframe computers. The documentation for UNIX (both its print library and its cryptic on-line commands and error messages) has been designed from a system-centered perspective that assumes more than a casual familiarity with UNIX . The major sections of Volume 1 of the UNIX documentation (see Figure 6.1) depict the system-centered focus on system features instead of user tasks.

1. Commands
2. System calls
3. Subroutines
4. Special files
5. File formats and conventions

Fig. 6.1.
Section Headings of a UNIX Manual

The documentation in the system-centered approach, as exemplified by the UNIX system, is a literal documenting of the static system: a description of the system's features removed from any context of use. The issue of where the computer will be used, who will use it, and for what purpose(s) it will be used is assumed in the system-centered view. A consequence of this approach for user documentation has been to place user documentation far from the center of system development (see Figure 6.2).

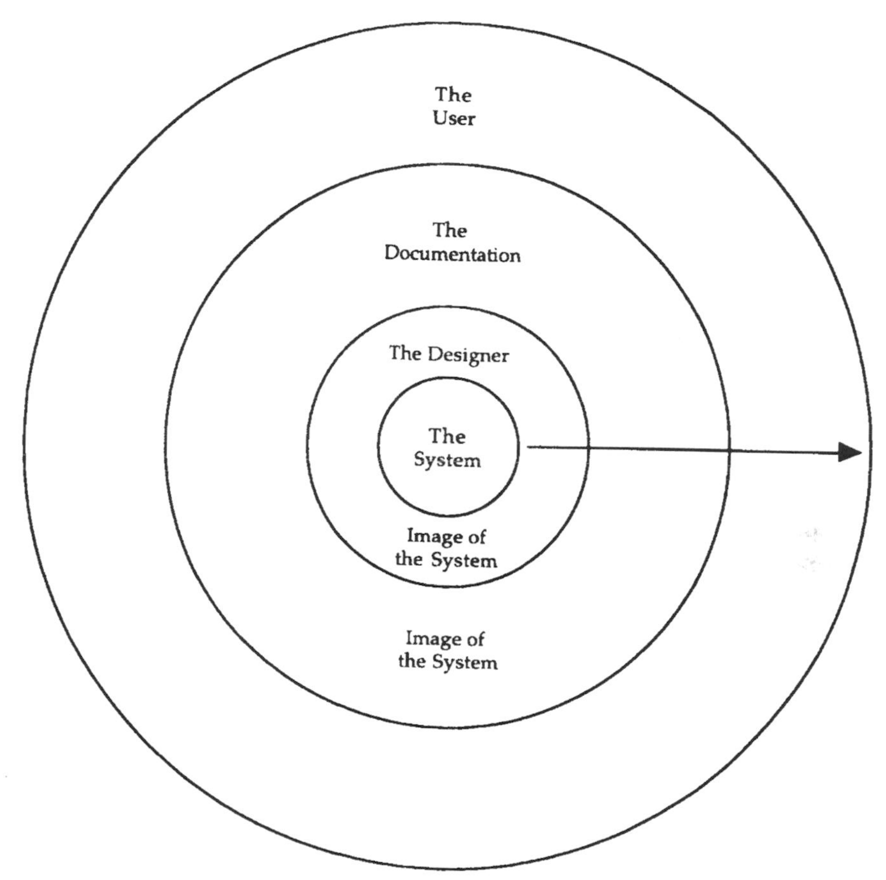

Fig. 6.2.
The System-Centered View of Computer Documentation

System-centered documentation places the needs of the technological system at the center and treats the system as the source of all knowledge pertaining to the development of documentation (as the arrow indicates). This view assumes that it is not only focal, but that the system exists before any documentation exists. There is no need for the user to have an impact on the central concerns of system development, this black box perspective of technology suggests, because the system is too complex and therefore should be designed and developed by experts who know what is most appropriate in the system design. The image that radiates from the system and eventually finds its way into the documentation, then, is an image that is envisioned by the system designer. Consequently, it is commonplace in system-centered computer designs to find cryptic error messages, long strings of difficult to memorize commands, and system components hidden from the interface that could prove potentially helpful to the uninitiated user.

From this designer's image follows the *documentation image of the system.* Here the documentation is written (often by the designers themselves, at least in draft form) to reflect what the designer views as the important components of the system. A common practice in this documentation development process is for writers to collect the "knowledge" of the sytem from the designers through notes written by the designers, or in interviews conducted by the writers. Consequently, the documentation is written to reflect the image of the system designer (refer to Figure 2.1 for an earlier description of the general system-centered process). The user, then, far removed from the central concerns of the system design, receives documentation that contains a system designer's image of the system.

I should explain that there are few published advocates of a system-centered approach to computer documentation, at least among professional writers.[8] Most technical communicators who write computer documentation are concerned with the needs of their audience—users—and therefore would not advocate a view of documentation development that places the user on the periphery. Unfortunately, technical communicators are often forced to be part of the system-centered approach because it is mandated (consciously or unconsciously) by the institution or organization they work within. Consequently, technical communicators write documentation at the end of the system development process and have little choice but to create system-centered documents.

Caught in a loop of technological determinism, system-centered documentation reflects the system designers' description of the system's features and its anticipated actions. The users' views of the system are irrelevant because they are not only figuratively but literally removed from the system development process. System-centered computer documentation has been viewed by developers as a necessary evil that must be created for the "idiots" who cannot figure out the system through their previous knowledge or trial and error. Put simply, system-centered documentation is

---

8. One published example of a system-centered approach to documentation is *Mastering Documentation* by Paula Bell and Charlotte Evans. Most system-centered documentation is produced in-house (and thus proprietary) with little or no published material explaining the process. The main evidence of the system-centered process remains the textual evidence it produces, like the UNIX manuals discussed previously.

the result of the deeply embedded dominance of system-centered philosophy in the world of computer technology development—a view that is far more technologically than humanly driven.

Texts, Readers, and "Reality": The User-Friendly View of Computer Documentation

User-friendly is a term that has been applied to technology usage in general (see chapter 2). In terms of computer technology, it refers to a computer product that has been designed as an attempt to fit the product to the background and expectations of the end user. Familiar icons on the screen to replace strings of commands, the advent of computer screens that look like the actual printed page ("what-you-see-is-what-you-get"), and the development of hand-operated pointer devices (i.e., the "mouse") to replace most actions done through a keyboard are a few examples of how computer interface designers have begun to make computers more accessible to a larger number of users.

The user-friendly approach to documentation development is characterized by an emphasis on the clarity of the verbal text, close attention to structured page design, copious use of visuals (often computer "screen shots"), and a warm, sometimes even excited tone that "invites" the user to enjoy learning the new computer system or software application. On a spectrum that places the system-centered computer documentation view on one end and user-centered views on the other, the user-friendly view would be in the middle, although in truth it is closer to the user-centered end. This view has more interest in the user than the system-centered view, and undoubtedly it has improved the quality of user documents to a great degree. The user-friendly view, however, still assumes that the design of computer systems is primarily the charge of the system designers and developers. That is, the system is assumed to be complete in the user-friendly approach, and user-friendly documentation is viewed as the vehicle for carrying the "reality" of the system image to the user. Therefore, user-friendly computer documentation has most often focused on the verbal and visual quality of texts that support the description of the system and the readability[9] of the text itself (see Figure 6.3).

The user-friendly approach is largely an outgrowth of the intensive work that has been carried out in recent years pertaining to document

---

9. I use the term *readability* in a strict sense, indicating the use of formal measures of determining the quality and comprehensibility of text, such as the Gunning Fog Index or Cloze tests.

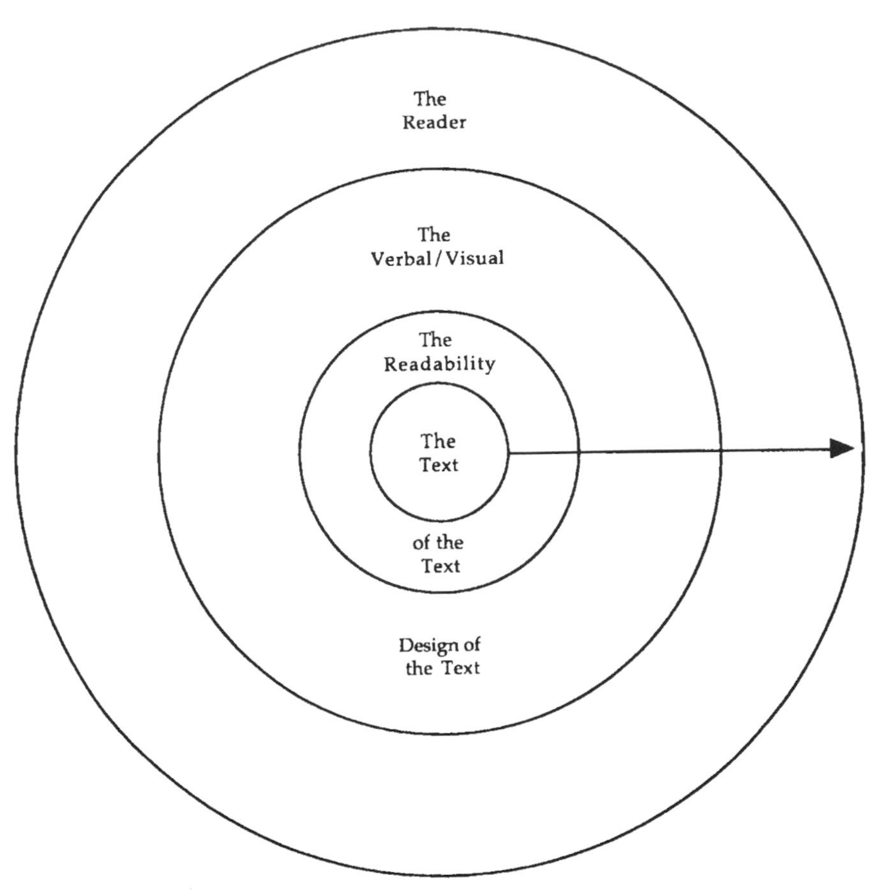

Fig. 6.3.
The User-Friendly View of Computer Documentation

design and readability. During the past two decades, document design researchers have argued that documentation should be designed in accordance with a process model of document design (Felker, 1981), and that the functional nature of the texts should be accounted for through careful structuring of their verbal and visual elements. From this perspective, the user-friendly development process begins approximately at the same stage of system development represented by the third ring of the system-centered view in Figure 6.2: the documentation, or text, image. Consequently, the user-friendly development process concentrates almost solely on text that has been written in accordance with the system designer's view of the system where there is little or no early user analysis before the documents are drafted. Most of these approaches also lean heavily on the traditional genres of user documentation, such as tutorials, user guides, reference guides, on-line help, computer-based training, and so on. This focus on text is clearly evident in most of the computer documentation textbooks currently on the market (Grimm 1984; Price 1993; Brockman 1991; Weiss 1991; Price and Korman 1993; Low, et. al. 1994).[10]

Radiating from this central concern of the text lies the readability of the text. The user-friendly approach emphasizes the importance of reader comprehension and learning and often advocates either readability measures or the results of previous readability tests to measure the validity of the texts that they are developing (see Duffy 1985 or Duin 1989). Great importance is placed on the coordination between the verbal and visual elements of user documents (as seen in the third ring of the user-friendly view, Figure 6.3). The integration of text and graphic is certainly not new to technical communication, as some of the earliest general technical writing textbooks spent considerable time on this practice (see, for instance, Mills and Walter 1978; Lannon 1997), and it has been the source of much important scholarly work in recent years (see Barton and Barton 1989; Tufte 1983; Killingsworth and Sanders 1988).

Finally, the fourth ring of the user-friendly view attempts to account for the user as a reader who is situated in a particular occasion of text usage: readers who have particular needs that can be fulfilled by documents designed to meet that occasion. Traditionally, the situation of the reader has been defined in terms of *learning* or *doing*,[11] and the texts that result from this interpretation of the user's situation fall into two categories: those that support learning (i.e., print genres of tutorials, user guides; on-line genres of computer-aided training or tours), and those that support doing (i.e., print genres of reference and quick reference; on-line genres like "Help").

The user-friendly approach has many advantages over the system-centered view, due to its focus on visual design and the needs of the user as reader: differences that have greatly improved the textual quality and user reception of computer user documents. The approach is a natural out-

---

10. I do not mean to imply that these texts completely ignore the user. Nearly all of the prominent how-to books for computer documentation writing include chapters devoted to methods of audience analysis or to brief discussions of user testing. The audience analysis methods, however, are usually of the "audience invoked" variety (see Ede and Lunsford 1984), which calls for a writer to imagine the needs of the user. User testing is most often presented in a late chapter as a way to test already drafted documents.

11. See Sticht ("Understanding Readers") and Redish ("Reading To Learn To Do").

come of the historical movement toward creating more usable texts. It does, however, have at least two chief drawbacks when applied to computer documents. To begin, because this approach is based on reading, it focuses on how well readers comprehend and follow printed text. Readability research has become closely linked to the quality of textbooks, and thus focuses on the comprehensibility, organization, coherence, and audience appropriateness of the texts' verbal content (see Duin 1989). This limits the results of such research to primarily the verbal elements of the text. Some researchers have expanded this view to encompass the use of visuals in user documents (Redish 1988), but most have concentrated on the reader's understanding of the verbal components (Charney, Reder, and Wells 1988). Such research yields valuable insights into reader behavior, yet it should be questioned in terms of how easily these findings can be transferred to the use of other media—primarily the computer screen. It is well known that people do not read computer screens as they do printed texts. Eye strain, impatience, poor resolution, and so on all play a role in the difficulties of reading the computer screen. In addition, users are not readers in the sense of using text in a linear fashion (i.e., reading from sentence to sentence, page to page, and chapter to chapter). Instead, users browse, access, skim, and jump from screen to print and back to screen again (Wright 1983; Sullivan and Flower 1986; Hartley 1985).

Secondly, user-friendly research points toward the active engagement between a reader and a text (see Charney, Reder, and Wells 1988), but it usually does so outside of the context of a user's actual situation of use. Even though the user-friendly approach attempts to account for the reader's situation (see the fourth ring of Figure 6.3), such research is often laboratory controlled for the purpose of collecting data that can be analyzed under the pressure of statistical significance. Laboratory controlled conditions cannot adequately reflect the complexity of a user's actual situation (see Duffy 1985), as the social contingencies brought about by organizational or cultural forces are rarely a focus of this type of research.

Let me reiterate that this is not a criticism of user-friendly research per se, because such things as readability measures for the specific purposes of determining comprehension and memorization of texts can be useful. Instead, I am questioning the findings that reader-centered, user-friendly research can bring to the problems of designing documentation for *user* needs. Users often act differently than readers (Wright 1989), and a theory of user-centered design should focus on how these differences rearrange our view of documentation. This means, for example, that the traditional guidelines regarding the verbal and visual integration of text might be redefined to meet the needs of the multimedia of user documentation. The medium can cause differences in the use of visuals, the format of the written components, and the quantity of information to be contained on each page or screen (see Bernhardt 1993).

To build upon and then move beyond user-friendly conceptions of user documents—toward a user-centered approach—means, in part, to complicate the documentation production process. Instead of looking at users as being merely active readers of text, user-centered design must ask questions of the user's situation, the medium of the documentation, and the organizational and cultural constraints placed upon the user and the documents. As I have already said, the research aims of the user-friendly perspective are driven by traditional conceptions of reading, and, as the following discussion of the user-centered view will emphasize once again, users are not just involved with the act of reading. Instead, they are involved in a complex of discourse that incorporates reading as one of a host of communication, social, and technology interactions.

Context, Negotiation, and the Medium: The User-Centered View of Computer Documentation

The user-centered view is philosophically and practically at the opposite end of the spectrum from the system-centered view. The philosophy of user-centered computer documentation places the user at the center, thus arguing for the user as the driving force in the process (see Figure 6.4). By placing users at the center of the approach, neither the text nor the system are dominant features as they were in the other views. Instead, the focus of documentation development is placed on the user. In addition, this view envisions the user as situated in a particular time and place: the user is not using the documentation to learn software abstractly, but rather is learning the computer application for a specific purpose or purposes.[12]

The core of the user-centered view, then, is the *localized* situation within which the user resides. For example, a computer user may be using a word processor to complete his or her work. The act of using the word processor, however, is not a localized activity but rather a generalized description of what the user is doing. A localized description of the user's situation attempts to explain the specific work-related activity that the user

---

12. I should emphasize here the approach that I am advocating is primarily suited to computer users who are conducting what Bødker would call "purposive human work" (*Through the Interface*, 1991). It also is akin to the philosophy of work-oriented design of computer artifacts as described by Ehn (1988) and others of the "Scandinavian Approach" to technology design (see chapter 4). Thus, the computer documentation I will focus on will primarily be for users involved in a work activity, not those using computers for entertainment purposes.

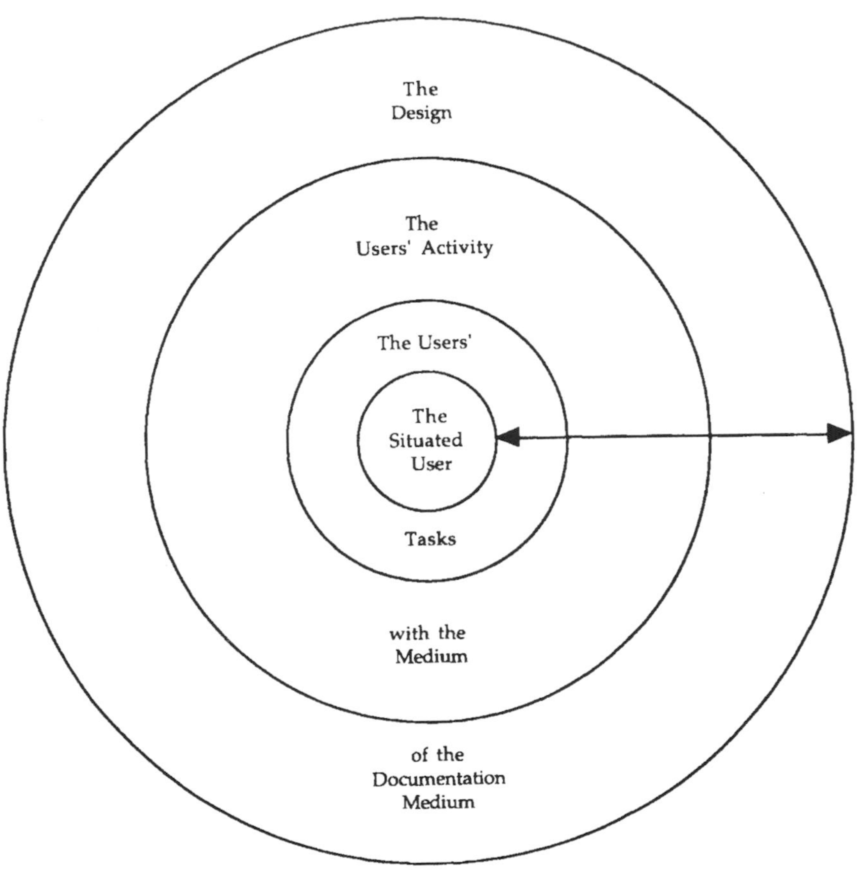

Fig. 6.4.
The User-Centered View of Computer Documentation

is involved with, such as the role that the user holds in an organization or institution. Thus, the localized user of a word processor would be described as a "bookkeeper," a "teacher," "a nurse," "a documentation writer," and so on. The specific nature of users' work, then, drives documentation that is customized to the context of use instead of generically explained to a universal user who is the construct of a writer's imagination. The focus on the users' situation also nudges the writer away from an immediate emphasis on the medium (which is still a very crucial component, as we shall see), and instead forces an analysis of the user as being central to the documentation process. This removes the writer from a premature preoccupation with the drafted document, as can be the case with readability or late-in-the-process usability tests where there is typically a relatively complete document in existence.

The user-centered view continues outward by taking into account the *tasks and actions* the user will be performing as a result of the users' situation (see the second ring of Figure 6.4). The focus on tasks within the user's context separates this approach from the system-centered approach, because the use of the computer is portrayed through user activity, as opposed to system features or system actions. A user-centered interpretation of a user's use of the word processor would be "writing a proposal," "laying out a page," or "producing a brochure." All of these activites could be considered "using a word processor," but the general activity depicted in the center of the user-centered scheme is too generic to accurately describe the actual activity of the user, especially concerning the documentation that will be written to help the user achieve his or her specific job. A more specific analysis of actual user tasks, as they are conducted within the context of use, is necessary.

I want to make clear that my concept of user tasks is considerably different than has been presented through *traditional task analysis* (see Bradford 1984; Weiss 1991; Barker 1998). In a traditional task analysis, a writer refigures the description of system actions into a vocabulary that is more relevant to the user's mental model of what the system is doing. Once a hiearchy of overall tasks has been outlined, writers delineate the subtasks that move the user from the start to the completion of the overall tasks. Traditional task analysis, which historically emerged about the same time as user-friendly approaches to computer documentation, was a significant improvement in documentation planning because it forced a user perspective into the system and software development processes.

The difficulty with traditional task analysis in terms of a user-centered perspective, however, is twofold. First, the user is relatively unsituated because, although traditional task analysis does move toward an understanding of user actions, it does so generically, not locally. Based upon a rational description of how the user *should* act, traditional task analysis merely reflects the anticipated actions of an idealized, logical user. The task analysis approach, in other words, is based on determining generalized knowledge that often ignores the local, domain knowledge of the actual, localized user.

Second, the tasks in traditional task analysis are still dictated by the system. The conception of traditional task analysis is embedded in a hyperrational approach to systems development: an approach that presents a completed system to the user with virtually no user input during any of the development phases. To be blunt, traditional task analysis is user-friendly, but certainly not user-centered. To avoid the rational replication of the system's tasks for the purpose of forcing the user to mimic them, the user-centered approach couches tasks within the situation of the user and seeks answers to questions like, What tasks will the user be performing within the given situation? Are these tasks truly user tasks, or are they couched within the terminology of system features? Are the tasks placed in an order appropriate to the situation of use?

In the user-centered approach, a close analysis of user situations and user tasks goes well beyond describing the rational system-dictated tasks expected of the user. Instead, the analysis attempts to understand the irrational or contingent occurrences that users experience within their local, everyday spaces. For instance, it is important in user-centered documentation to illuminate fundamental characteristics of users' situations to describe those *cunning* solutions that users have developed for dealing with technology. User knowledge of this sort was discussed in chapter 3 in terms of *metis*, the ancient Greek conception of the cunning intelligence that users display in everyday contexts. In a similar vein, feminist sociologists have termed this type of knowledge "articulation work," and as Susan Leigh Star explains, it is a type of knowledge that can only be "captured" through observation of users in their everyday environments, or direct discussion with them in that everyday experience.

> Articulation work is the work of fitting things together in real time in the workplace. . . . [For example, in a medical setting] Articulation work may mean tending to a patient, repairing a sentimental mistake in a conversation with a family, or finding a substitute implement in the middle of an emergency surgery. You cannot predict ahead of time what might be needed because, by definition, articulation work is contingent and occurs in "real time." Some people's daily work, such as that of secretaries, homeless people, and parents, seems especially rich in articulation. . . . " (Epilogue, ST&HV, p. 504)

To reveal these moments of user knowledge is crucial to documentation. It "gets at" the essence of human involvement with technology in a situated way that describes the productive nature of user knowledge within context. These moments of *metis* or *articulation work* depict users producing knowledge, or at least displaying knowledge that they themselves have constructed/produced in the past and are now using to perform in the present situation. Such localized, domain knowledge is unaccounted for through most computer documentation development processes, and, subsequently, the localized cunning knowledge of the work environment fails to surface in the written texts themselves.

Beyond the task ring represented in Figure 6.4, before the medium of the documentation begins to play a central role in the design of the documentation, is the *users' activity with the medium.* One of the most complicated aspects of computer documentation specifically, and of instructional documents, generally, is that they are *used in conjunction with the act of using a technological artifact or system.* Consequently, it is important in user-centered design to determine which medium will best fit the particular user situation and tasks. For computers, this distinction has become especially relevant since the advent of on-line information and has been the source of numerous questions for documentation writers. Whether the documentation should be delivered in print or on-line is a question asked most often in discussions of documentation development. Will the user be better served by a print document, on-line document, or both? Will the user be using the medium in a linear or random manner? Is the medium capable of providing the user with a familiar model? For example, are the on-line system capabilities of the system too embedded within the interface, and thus users would be better served by the more familiar medium of print?

In addition to a choice of medium, the type of activity that the user is engaged in must be assessed. In the Discourse Complex of User-Centered Technology (see chapter 2) the three general categories of activity were described as *doing, learning,* and *producing.* In the context of computer documentation, *doing* describes activities where users are not reflecting[13] upon their actions for the sake of long-term retention. For instance, intermittent users of a certain piece of software may just want to be "refreshed" concerning the procedures of an action, but they will have little reason to retain that information because they rarely perform the task. Instructions that support the role of *doing,* then, usually will be less elaborative, more action-centered, and highly specific concerning the nature of the users' situation.

The term *learning,* in the computer documentation realm, is revised to *learning through doing.* This change in terminology accounts for the problems associated with learning about computers while simultaneously using

13. The term *reflection* as used here is not meant in any way to indicate that users are unreflective, or that the ethics that might be involved with reflection on one's actions is absent from such a scenario. I merely mean that users are doing tasks that must be accomplished and that there is little time for memory retention. A good example of this are the pull-down menus of computer interfaces. People often cannot remember under which menu item a particular command exists, but once they are placed into a context of use (e.g., they are actually using a computer interface) they can immediately go to the correct menu choice. They have not taken the time to learn the actual placement of certain commands, but they nevertheless are knowledgeable users of the system.

the computer: a paradoxical situation where you are compelled to learn (maybe because your job depends upon the new technology), but you are actually more interested in just completing the activity at hand.[14] This complicated paradox of learning and doing is, nevertheless, a real problem for users, especially those who are experiencing electronic technologies as they emerge into existing domains, such as the workplace or classroom. For example, if a medical technician is using a computer for the first time to analyze results of a blood test, then the technician is bringing expert knowledge of one domain (blood test analysis) to a new domain (computer technology) where the technician has little or no knowledge. For a technical writer who is determining what to include in the documentation for the blood test analysis software, it can be unclear which domain of knowledge is affecting any breakdowns that might occur.

More importantly, the user (the medical technician in the aforementioned case) is learning to carry out tasks *at the same time* he or she is doing the tasks. The scenario of this activity would be that the user is, at one moment, using the artifact (the computer), "reading"[15] the documentation (print or on-line), and attempting to extrapolate data that is meaningful in his or her knowledge domain. To account for this complex of activity, I have expanded the term learning to *learning through doing* to better represent the activity that users are engaged in while learning computer software. Documentation is compelled to manage this dichotomy for users through the careful design of print and on-line documents that can accomodate learning *and* doing simultaneously.

Producing describes two specific activites of users as they are involved with documentation development: activities that are not accounted for in either the system-centered or user-friendly views because neither approach invites actual users into the development process. First of all, users in a user-centered appraoch actually take part in the production of the documentation. They are involved in such things as initial design planning, iterative evaluation and testing of the documentation or software, and finally the decision making concerning the implementation of the documentation in

---

14. Mary Beth Rosson and John Carroll refer to this as "the paradox of the active user." Their observations of users showed that users want to learn, but often are more interested in just "getting on with it."

15. Reading is in quotation marks to designate the problems of reading discussed earlier in terms of the user-friendly view. I contend that computer users do not read in any conventional way, and thus the concept of reading is problematic. This is true if we are designing documentation that we anticipate as material to be read as if it were an "armchair" document or a textbook that one reads in a library carrell, removed from the context of technological use.

their respective contexts of use. Second, users are producers in the sense of knowledge production (see chapter 3). What they know about the technology in question—its documentation, interface characteristics, and situations of use—are necessary prerequisites for involving users in technology development and decision making. The user's physical presence in technology development is not the only thing of importance; the knowledge of the user must be valued as well. In a user-centered approach to documentation, the *producing* user is crucial because the process of user-centered documents is a collaborative, negotiated affair that is perpetually building the knowledge of users and their various contexts into the products being developed.

In itself, the social interaction of users, developers, writers, and marketing specialists should be adequate reason for adopting user-centered approaches to many different forms of technology development, including documentation. Unfortunately, the act of involving users so intimately in the development process is met with much resistance. One barrier is the already mentioned schism between experts and novices. Users, because of their role as the "unknowing subject," are deemed unworthy of participation in the highly specialized world of the technology developer. More likely, however, the reason for noninvolvement of users is the perception that it simply is too costly. Such arguments concerning profit margins have merit, but only within the context of short-term economic strategies. Certainly it is more expensive to commit time and resources to user involvement on a per project basis, especially if each project is seen as being unconnected to future versions of, or extensions of, the product. User input, though, is something that has value far beyond the present moment of testing a particular product. Capturing the knowledge of users and tracking such information over the long term could only be beneficial.

The long-term advantages of allowing users to play a role in the production process are potentially great, however, and some research has pointed to the positive effects for financial reward and customer satisfaction when actual users are allowed to work with the production teams (Ehn 1988; Nielsen 1993). Full effect of user involvement in production, though, has never been measured because of our culture's proclivity to base most production decisions on short-term motives and outdated cost accounting procedures that focus, in a very limited way, on month-to-month (and sometimes even day-to-day) analyses of the "bottom line" (see Johnson and Kaplan 1987). User-centered approaches to technology development are, in part, counter to the short-term mind-set of business planning because the benefit of user involvement can best be measured over the long term.

At the outer ring of the user-centered view is the *design of the documentation.* User-centered designing of the medium is now possible because the prerequisites of the user's situation, tasks to be performed, and purposes for using the chosen medium have been determined. The placing of the design or writing of the documentation at the outside of the user-centered view should not be taken perjoratively, however. Rather, it is the end of a production process that is consciously driven by the needs of the user. This does not diminish the role of documentation. On the contrary, it is an enhancement because it argues for user documentation that is in the image of the user's model, not the system model or system designer's model. A user-centered approach, in essence, is a thorough form of audience analysis that is aimed at designing documentation that fits what a user *actually* does, not necessarily what we *think* he or she should do.

A final feature in the user-centered view is represented by the arrows that indicate movement in both directions, instead of the singular inward to outward movement of the other two perspectives. In the system-centered and user-friendly views, knowledge is put forth from the place of a singular authority (either the system or a text). These nonnegotiable views not only treat the user as a mere receiver, but they also exclude the user from an active involvement in the processes of production, as mentioned earlier. The two-way movement of the user-centered view depicts the recursive and negotiated nature of the user-centered approach to computer documentation, specifically, and technology, generally. The user-centered perspective invites involvement by the user throughout the process. The two-way movement attempts to disperse authority through a recursive process that is always in motion and always correcting itself, dependent upon situational contingencies. In an actual documentation production cycle, this recursive movement from user out to the medium and back again could occur numerous times—a truly iterative design technique.

To further elaborate the user-centered perspective, I will now turn to the user-centered rhetorical complex of technology (see chapter 2) for the purpose of revising it to meet the context of computer documentation (see Figure 6.5). I will begin by describing the major changes that alter the complex for the context of computer documentation.

The Rhetorical Complex and Computer Documentation

I do not intend to fully elaborate upon the entire complex here (for that you can turn to chapter 2). Instead, I will focus on either those elements of the complex that have been renamed for the purposes of defining user-centered computer documentation, or on those that need additional explanation because of their role in the processes of computer documentation production.

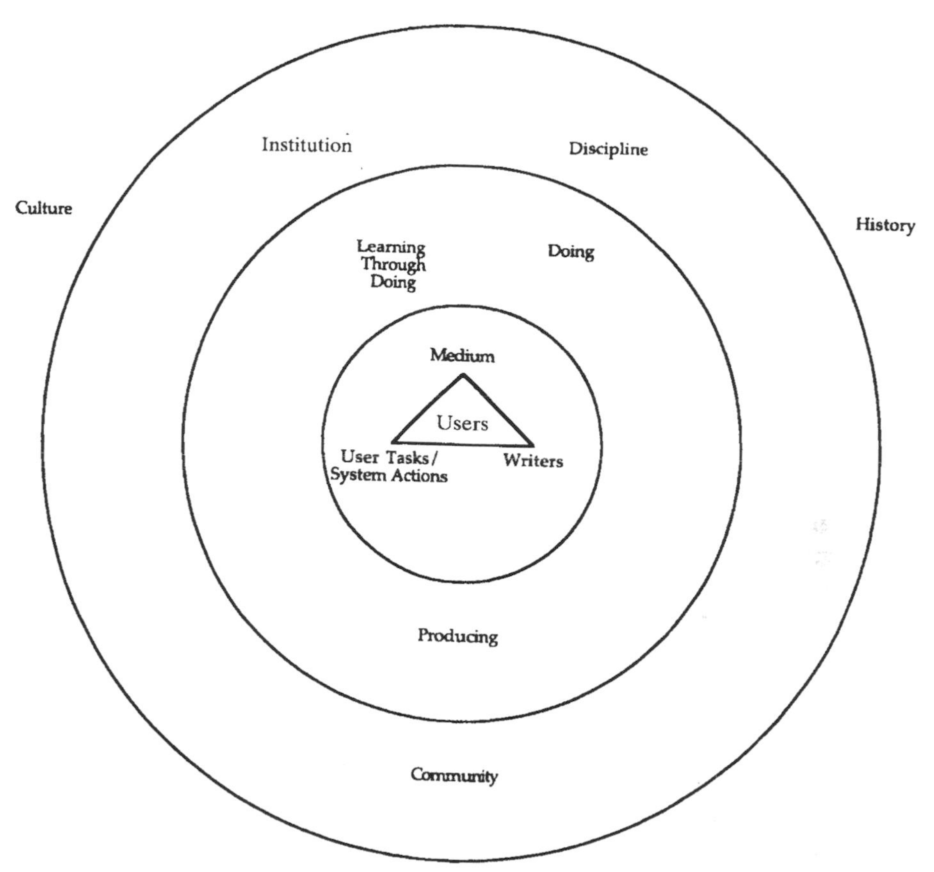

Fig. 6.5.
The User-Centered Rhetorical Complex of Computer Documentation

The artisans/designers of computer documentation are the *writers* who plan, design, and produce the user documents. There have been many discussions over the years pertaining to the label of "writers" as a meaningful designation for those individuals who create the instructional texts for computers and other complex technologies (see Haselkorn 1988). Companies have termed the producers of documents as information developers, document designers, communication specialists, or any number of other titles. The Council of Programs in Technical and Scientific Communication has carefully chosen the term *communicator,* in part to clarify that technical communicators do more than just work with written text. I do not wish to debate the relevance of accurately describing the activities that technical communicators do in their jobs: technical writers clearly do more than just write text. Nevertheless, given the potential shortcomings of the term, I have chosen *writers* to designate those who plan, design, and produce the various types of texts we call user documentation. Thus, writers become the artisans/designers in the complex in the context of documentation development.

The artifact/system is changed to *medium* to reflect the nature of the artifact that the writers work with. Documentation writers may be writing about computers, but the artifacts they create are the documents that reside in different media. Documentation, of course, comes in two primary media—print and on-line—but there are numerous variations of these media as a result of video, audio, animation, color, voice, telecommunications, and hypertext structures. Consequently, what was once the relatively "simple" and stable artifact of print is now much more complex, at least from a writer's point of view. Also, the advent of desktop publishing has made the role of the writer much more diverse than just a mere scribe or wordsmith who transfers the information about the system from the *system designers* to a text. Those writers who work with print technology are expected to produce camera-ready copy, a phenomenon that was rarely seen before the development of the laser printer and postscript technology. The medium, then, is not only an important factor for the user of technology but for writers who must meet the challenges of learning to "write" for different communication forms.

The terminology of the third point of the triangle, the *user tasks/system actions,* does not change literally, but the characteristics of this feature are of utmost importance in computer documentation and thus deserve some further explanation. As explained in chapter 2, the *user tasks* are the actions of use as perceived by the user. The *system actions,* on the other hand, are the actions that the computer system was programmed to accomplish by a system developer. The point at which these two aspects of technology meet is commonly called the *interface*: the place where the user meets the machine during the context of use.

The disjunction between user tasks and system actions is at the heart of the problem of user breakdown. Users often perceive their use of technology differently than the represented actions of the system. For instance, in fax machine documentation the action of sending a message to multiple recipients is commonly called "polling." While this terminology might be clear to the developer of the system, it is the rare user who perceives the action of sending to multiple recipients as "polling." Users would perceive this task as "sending to multiple recipients," or "distributing to more than one client." It is this visual and verbal description of the system (in print, on-line, or as a label on a control panel) that represents the use of the technology that is meant by the rhetorical element of user tasks/system actions.

For writers, the crucial problem concerning user tasks and system actions is to understand the nature of these breakdowns and adjust the documentation (in the best of worlds, to fix the system itself) to represent the users' perception of tasks.

The aforementioned descriptions of user-centered concepts should offer some insight into the possibilities for application of the concepts to actual documentation writing practices. In any case, they offer a foundation for further discussion of what the impact of user-centered theory might be on documentation development practices. In the next section, I will use this foundation to discuss the issue of documentation genres—the types of documents that are written for users of computers—to investigate the nature of documentation within the rhetorical complex.

Genres in the Making: The Case of the Ubiquitous Tutorial

Genres are taxonomic devices that provide order and meaning to many everyday artifacts. While genres can be used to categorize any number of artifacts, they have been used widely to categorize virtually all types of text.[16] When we pick up a newspaper, for example, we immediately have certain expectations about what we will be reading, how we will move through the text, and what we will do with the text once we have finished with it (wrap fish, of course). We know the category of text (journalistic) and what subcategory of journalistic text it is (a newspaper). This tacit knowledge of text places it into groupings that aid us as we interpret it, and use it, in particular contexts.

Genres like newspapers have been around for quite some time: they have a historically established context of use, and the expectations for these genres is generally understood by a wide audience. When traditional genres are resituated within a new context, though, the revised use and purpose of the text can be the cause of a reformulation of the genre itself. This is currently happening with the genre of newspapers as they are put on-line over the WorldWide Web (although I hesitate to guess what the new genre might be called because the phenomenon is so new). The generic structure and purpose is fluid and being redefined as people learn how to use, and how they want to use, the new on-line medium of journalism. Thus, the situations we use texts within, and the technological medium of the text, have

---

16. For more on genre, see chapter 3 of JoAnne Yates' *Control Through Communication.*

a great deal to do with how they are used and how they should be designed. In a discussion of technological artifacts as genres, John Seely Brown and Paul Duguid offer the following example of this phenomenon.

> No artifact is self-sufficient. The spot-lit, pristine artifact commanding the center of attention among the usual array of potted plants at a trade show reveals very little about whether, or how, or why it will or will not be used. But the artifact in the workplace, plastered with stick-ums or Scotch tape, modified or marginilized by practice, and embedded in social activities, can tell a rich, well-situated story. Consequently, in contemplating usability, designers cannot just consider isolated (and isolating) notions of functionality. Instead, they have to relate these to the socially and physically embedded practices. (p. 174)

Brown and Duguid go on to explain that in the case of genres, even though they may appear generally static, they are continually changing. "The underlying generic conventions, though well established and remarkably widespread, are not predetermined. They evolve locally and continually in practice. Change is prompted from many different directions" (p. 177). Practice, or use, is a determining factor in the development and recognition of genres. It is not the only determinant, but certainly the context of use frames genres in certain ways that make text recognizable to different users who are bound within a common situation, such as a community of users in a place of work.

Computer documentation genres are an interesting case to investigate concerning fluid, evolving genres. Documentation, either print or on-line, certainly is constrained within situations of use, and, as a genre, computer documentation should be perceived as mutable in terms of how it is designed and constructed. In addition, the user, as the primary agent of practice or use, also should be perceived as an agent in documentation design. In reality, however, the genres of computer documentation are rarely so flexible. Computer documentation historically has been designed to conform to predetermined notions of users and use: notions that come from a time when computers were designed for use by experts, or as we have seen in earlier chapters, for idiots. Thus, computer documentation is driven by preconceived ideas of what they should contain, even if those contents have little to do with the intended audience's needs.

As one peruses most general technical communication textbooks, or texts written specifically for the writing of computer documentation, it is evident that the genres of computer documentation (print or on-line) generally are broken into two overall categories—*texts for doing and texts for learning*. The *doing* texts usually are suggested for the advanced, or expert, user and include print documents such as reference guides, quick reference guides, and troubleshooting guides. The on-line forms of the *doing* texts are loosely placed in the large category known as on-line assistance or "Help." Interestingly, the *doing* texts are found in a somewhat wide variety of shapes, sizes, and other design features. For instance, quick reference information has long been customized to meet the need of users ranging from computer users in an office setting to engineers using pocket-sized guides for information needed "in the field" (see Reitman 1988).

The *learning* texts, however, are not so flexible in structure. These documents—although subcategorized under labels like user guides, guided tours, and computer-based training—are in most circumstances generically termed *tutorial*. The genre *tutorial* is supposedly well defined as a "use once, throw away" document—a no-deposit, no-return text that is supposed to get the user "up and running." Theoretically, the tutorial will be used in a linear, step-by-step manner as the user moves from one lesson to the next, each lesson presumably slightly more difficult or advanced than the previous one. Once the user has completed the assigned lessons in the tutorial, it is assumed that the user will abandon the tutorial and proceed either on his or her own volition, or with the occassional assistance of a reference guide.

In practice, however, tutorials are seldom used once and then abandoned. In the worst case, the tutorial is the computer world's epitome of the "when in doubt, use the instructions" attitude that began this chapter. In these cases, it is more common than not that the tutorial will remain in the protective shrink wrap, not to be used at all.

In the best case, the tutorial can become a long-term resource for the user. Because the user has become familiar with the structure, look, and language of the tutorial, he or she returns to it often to refresh the memory, or maybe to learn from a lesson that was never completed because either time ran short or the relevance of the lesson was unclear at that moment. Tutorials also can appear less intimidating because they often are shorter and less bulky than the large reference documents that portend to contain all of the knowledge of the system. Unfortunately, tutorials that are consciously designed to support long-term, user-centered goals are far less common than those designed for one-time use. In fact, the tutorials that are used in the long term probably were not consciously designed for that purpose. Instead they have been adapted serendipitously for long-term use by the user with techniques that use sticky notes, highlighter pens, and dog-eared page corners.

There is no reason, however, that tutorial documents could not perform a broader array of functions for users. First, however, the genre "tutorial" will have to be redefined. The concept of tutorials as *learning through doing* documents offers one perspective on how this can be accomplished.

If learning documents such as the ubiquitous tutorial are re-thought through the problems of synchronous learning and doing—learning through doing—then they can become a more customized text in terms of design. The process of creating such documents, of course, entails analyses in accordance with the user-centered view (see Figure 6.4): the user's particular situation, specific tasks, and proclivity for and access to print and online media would be necessary. While I will not carry out such a process in detail here, I will use an example from a user-centered document that in many ways overcomes some of the difficulties of the use once, throwaway tutorial (See Figure 6.6).

Figure 6.6 is from a manual intended to introduce a new user to a database software (File Maker Pro® 2.1). The specific user of this document is a secretary who has the responsibility of keeping the records of applicants and current students in a graduate program at Miami University. The secretary's computer experience had for many years consisted of using an early Wang® system that had no graphical capabilities and limited word processing capabilities (in other words, it was not a what-you-see-is-what-you-get system). This particular secretary, however, was highly experienced with the context of this office setting, as she had held the position for nearly a decade before the new Macintosh® system was introduced.

As you can see from the two pages excerpted from the manual, the user's situation is central as the document has been customized to focus on just the tasks associated with applicants to, or current students of, the graduate programs. Thus, specific instances of entering information appropriate to this context are used in the examples of the manual. On the first page of Figure 6.6, the italicized information immediately to the left of the computer screen "shot" indicates how to size a box of "Degree Sought" information. This direct link to the user's specific context reduces ambiguity pertaining to the uses of particular fields in the database; it also creates a familiar context within which the user can immediately learn an action while actually accomplishing a necessary task related to her work.

Also notice that the information of "how-to" is explained in more than one way. At the top of the second page in Figure 6.6, for instance, the user is given three reasons for changing the layout of the screen. These three suggestions are then followed by an explanatory note that further justifies some of the choices for layout arrangements. It could be argued that such choices might interfere with a user's notions of completing a given task because the choices could lead to confusion on the user's part. In the case of this manual, however, the writers observed that she specifically wanted information that could be used more than once: she needed the document to be an asset in her different stages of learning the database.

Changing the Format of a Layout

Each arrangement of Graduate Program Applicants information is called a layout. You can easily create a new arrangement of information or change the way information is displayed or treated in fields by creating a new layout or editing an existing one.

To create or edit layouts, change to the *Layout* mode.

Getting to Layout Mode

Get to *Layout* mode by

- choosing it under the **Select** menu

  or

- clicking the mode button to the left of the bottom scroll bar to bring up a pop-up menu. Select LAYOUT from this menu.

Layout mode looks like *Browse* mode, except you get rulers, section breaks, and editing tools for changing the size, appearance, and placement of fields.

The square corners around the "Degree Sought" field indicate that this field has just been selected. It may now be moved or resized with the cursor, or its appearance may be changed by selecting options from the pull-down menus.

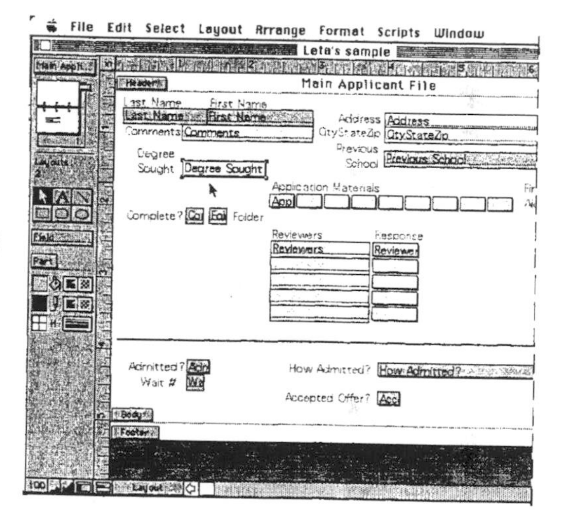

For On-Line help see

Designing Layouts

For help from manual see

Working with Layout Parts and Fields 3-27 to 3-59; Adding, Editing, and Formatting Layout Text 3-60; Moving Objects 3-82

Fig. 6.6.
Two Pages from the Database Document

Changing the Arrangement of a Layout

You may want to change a layout for several reasons. You may want to adjust it so it's easier to read on the screen, you may need to add or delete a field, or you may need to set up a layout for printing different versions of a report.

NOTE: It seems to work best to have separate layouts for viewing and printing, since features like colored boxes and field borders make reading information on the screen easier, but significantly clutter up a printout. Also, some fonts and font sizes will be more readable on the screen than others, but they may be too big or less readable when printed. Separate layouts for viewing and printing allow you to control the appearance and of your text with formats you set up once and use over and over. (For an example of a good print layout, see the section on compiling a list of applicants by program, p. 12.)

In *Layout* mode, items in a layout may be moved around simply by clicking and dragging on them with the pointer tool (the arrow). Use rulers, and grids to help guide placement.

Changing the Appearance of Text

To change the appearance of text in the layout

- Use the text tool (the A box) to select or highlight the text you want to change. Select the new appearance from the options in the **Format** menu.
- Use the pointer tool (the arrow) to select the fields you wish to change. Each selected field will be highlighted by small boxes in the corner. Select the new appearance from the options in the **Format** menu.
  (Holding down the shift key while selecting fields allows you to change more than one field at a time. You may also SELECT ALL fields on the **Edit** menu.)

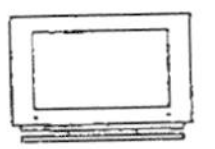

For On-Line help see

Designing Layouts

For help from manual see

Adding, Editing, and Formatting Layout Text 3-60; Changing the Shape and Size of Objects 3-82;

Fig. 6.6. (continued)
Two Pages from the Database Document

Another user-oriented characteristic of this manual is the use of the two icons at the bottom of each page—one depicting a computer screen, the other a printed text. During observations of the secretary carrying out her work, and through interviews with her in the design process, the writers discovered that she never turned to the database's on-line "Help" for assistance. The old Wang® system did not have on-line Help, and consequently she had never witnessed on-line assistance in any of her previous experience with computers.

Through further analyses of the secretary and her use of on-line Help, the writers defined two problems related to the use of different media. First, they observed through user testing that when she was prompted by them to use Help she appreciated, to some extent, that the information was "at her fingertips." Despite this appreciation for the expedient nature of the on-line documentation, her proclivity was to use print documents when she needed assistance. To provide assistance to her that would account for this proclivity, the writers devised the print and on-line options at the bottom of the pages. They additionally felt that the icons would be highly visible and that, eventually, she might begin to use the on-line Help information with greater frequency if she was reminded of its presence through the icons on the printed pages of the manual.

The icons at the bottom of the pages also serve the purpose of providing the user with immediate feedback concerning the whereabouts of further instructional elaboration. As you might expect, this customized, situation-specific manual would not be expected to contain the extensive "complete" elaboration found in the commercial documentation accompanying the software package. To overcome this potential deficiency, page number references and topic identifiers were added adjacent to the icons. Through the page number references and topic identifiers taken from the industry-produced publications, the writers provided access to greater amounts and varieties of information: a feature that produces something of a "hyper" quality to the documentation as the user can move in random, yet linked, ways though the entire library of documentation for the software. Thus customized reference by specific page number and topic allows for integration of the entire set of user documentation for this database software—an integration guided by the actual situation of use.

The writers also found some interesting vocabulary problems that may very well have been the result of an expert-driven task analysis procedure, as discussed earlier in this chapter regarding user-friendly documentation. While investigating the problem of integrating the two mediums of documentation, the writers uncovered inconsistencies between the respective vocabularies of the existing print and on-line documentation for the database software. As they collected the task vocabulary for both the print and on-line information, they encountered quite different terminology for the different tasks. For instance, at the bottom of the first page of Figure 6.6, the on-line terminology is "Designing Layouts," but the print documents refer to this task as "Working with Layout Parts and Fields," or "Moving Objects." Instead of using one term for each task, the writers decided to offer the various renditions of the task vocabulary as they appeared in the different pieces of the documentation set. This feature offered a mini-index on the same page as the how-to "learning" information, making the "tutorial" a quasi-quick reference manual too.

It is somewhat easy to get involved with the analysis of texts and forget that there might be other issues at stake when we think about instructional documents. The title of this chapter, I believe, should be a constant reminder to technical writers that users have little affinity for instructional assistance: they will use it only when all else fails. Put another way, users of most technologies, and most pointedly users of computer technologies, often see documentation as something that can get in the way. The very presence of documents, whether print or on-line, presents the danger of users becoming disengaged from the learning/doing processes, because the documents can draw the attention away from the activity at hand.

An obvious consequence of this propensity by users to avoid instructions is that documentation, and those who write it, will remain invisible in the technology development and implementation processes. To a great extent this is already the case, as the perceived need for on-line texts is somewhat dubious. For example, I would argue that the tendency for some software manufacturers to put all of their products' documentation into on-line formats is driven more by economic concerns than by concerns for the users of the documents. Some on-line documentation is nothing more than the print documentation reentered into on-line Help with little or no attempt to alter the information to fit the use of the new medium. Such lack of interest in the design and use of on-line documentation, although not a new phenomenon, certainly does not bode well for its future as an important, viable component of everyday user technologies.

The invisibility of writers also is a problem as they are hardly at the top of the technology development hierarchy. Too often the role of writers has been "to write it up" at the end of the development process: a problem that clearly is a source of the system-centered, functionality-based documentation and interface designs that have dominated the personal computer industry for most of its short but influential history. In brief, the issue of empowering writers within their workplace contexts looms large in the domain of instructional text. In the conclusion of this chapter I will take up these two issues of invisibility regarding technical writers and the texts that they produce.

To Document or Not to Document?

As we already have seen in several instances, users face a double bind when they are confronted with the problem of learning new technologies through instructional texts. To engage with a technical artifact *and* a text at the same moment is a complex and frustrating task that illuminates the paradox of learning through doing. In their research on the minimal manual, John Carroll and his associates characterize this paradox in a way that is most useful in clarifying this fundamental difficulty of technological use. Carroll et al. differentiate between two types of learning procedures: *learning by the book*, and *learning by doing*.

> Learning by the book models the kind of learning process for which self-instruction manuals as a genre are designed. Learning while doing attempts to model procedurally the task-oriented skipping and self-initiated approach . . . which we have taken to be characteristic of real learning. . . . (Carroll 1990, p. 176)

In the previous discussion of the secretary's database, we saw how learning by the book and learning by doing might be confronted through documentation that challenges traditional notions of genre. The lenses used to design the new genres, however, may be rose colored; it can be argued that the presence of documents in the context of active learning is itself futile. Will users even use these documents that have been carefully crafted for their contexts? If texts get in the way of active learning, this skeptical perspective might ask, why create user documents at all? Why not just make the interface so usable that documentation is never needed?[17]

HyperCard® for the Apple Macintosh® is an example of this perspective that advocates the dissolution of documentation. Especially in its early days, Hypercard® was touted as being so easy to use, due to the graphical interface of the Macintosh® and the icon-driven HyperCard® application, that the documentation for learning Hypercard® was merely created as adjunct material hardly needed once a user had a very brief tutorial introduction to the application. Anecdotal evidence of this attitude toward

---

17. This certainly is not a novel claim, as the eradication of user manuals through better interfaces has long been the goal of interface designers like Donald Norman.

HyperCard® can be given, but the following statement from a HyperCard® User's Guide provides some evidence:

> This ends your tutorials on HyperCard's basics. You know enough now to get off to a great start. You can go on reading in this manual to learn more about HyperCard or you can go off on your own, browsing and editing as you like. . . . Happy wandering! (p. 59–60)

One of the concepts of HyperCard®, and hypertext in general, is that it is a user-controlled environment, and therefore exploration, play, and experimentation are activities users can engage in to learn. Such a concept can be good. If a user is enjoying himself or herself, then it is probable that learning will occur. The depth of the learning, and the relevance for that style of learning within a particular context, though, is problematic. Play and experimentation would be fine when there is little pressure on the user to perform, but in a workplace or educational context some guidance and strategies for use probably are more appropriate. In the case of Hypercard®, for example, other levels of the software are useful, even necessary, for users (i.e., scripting, importing and exporting graphics or video) that would be difficult or nearly impossible to learn without some documented advice or guidance.

It does not seem at all clear to me that computer documentation will disappear. But we can be assured that the role, look, and medium of documentation will shift. It may even be the case that computer documentation will have to blend into ill-defined contexts, like those of intense collaboration between users as they learn networked electronic technologies. To ensure that these shifts are user-centered will take a concerted effort by writers to assure that it is not a determined shift: a shift dictated by the technological imperative that we hear so often in phrases like, "It just cannot be helped. This is the world we live in now!" In short, we should be ready to help define any shifting before the shifting pressures us into documentation practices that neglect user-centered goals.

Beyond the Text: Writers, Writers, Everywhere

As the discussion of shifting genres demonstrates, the user-centered argument concerning computer documentation is not without roadblocks. There is one issue, however, that may be the most problematic of all—creating customized documentation for specific, localized contexts and the corresponding number of writers needed to design and develop the documents. Arguments for situating documents within contexts are not entirely new. Literacy advocates, for instance, have argued that literacy training is much more effective if situated within contexts relevant to the learner (Sticht 1985). This makes sense. Many people are more motivated if they can see a connection between what they are learning and the intended application of their new knowledge. Unfortunately, putting such localized ventures into practice is more easily said than done.

A most obvious difficulty is that writers who are knowledgeable in user-centered design must be hired to create these documents. This not only means that software development companies might hire these writers, but more specifically, it argues for institutions using the software to employ writers who can, on an ongoing basis, create documentation for the constantly fluctuating audience of users in the institution's employ. At present, such a scenario—where writers would be hired to create numerous custom documents—might appear absurd. How would institutions that are facing downsizing and other cost-cutting pressures make room for writers? There is no easy answer to this. But let me attempt to explain.

The usability evaluation and testing that must accompany user-centered design is costly, but, as I have already mentioned, only in terms of short-term management. These costs can be recovered over time through better design decisions that hopefully make the products more accessible to a wider body of users. This argument has already been made in several quarters in terms of "discount" usability (see Nielsen 1993). Most of these solutions, however, do not include technical writers. To include writers will mean efforts to recast writers. Writers will not only need to change the moniker that denotes them, but they will need to find avenues into other domains of technological design and development. For example, I have observed institutions that regularly increase the number of system administrators as computer usage increases within the organization. These administrators, in turn, take on various roles that include user liaison or software support. Writers could easily fulfill many of the functions of these system-trained administrators, particularly those tasks that involve the learning of systems. Companies and institutions that recast writers in new roles such as these could also benefit from the development of the print and on-line texts the writers would produce. The information about how to use the software would no longer be in oral (word of mouth) format, but instead would be formally documented, thus making documentation updates and more widespread, consistent training possible.

In addition, if institutions learn to value the worth of training materials more highly, especially materials customized to fit specific user situations, then writers will have more central roles in the maintenance of user systems. Surely many institutions have users who regularly underuse the computers in their domain because they are either unaware of the potential or are too strapped for the time to learn new features that would aid them in their jobs. Consistently updated custom documentation would be a great asset in such situations.

As documentation increasingly goes on-line, the role of writers should increase as well. Writers will certainly be called upon to enter the old print information into on-line formats, and this should be done, of course, through user-centered methods if there is any intention of creating usable products. In addition, writers should increasingly take on the role of interface designer. The interface, after all, is a "text" that resides between the user and the black box of the system—a multiple media text that includes documentation. Writers already are interface designers, in other words, but recognition of this technical communicator skill is rare outside the realm of technical communication.

A recasting of the technical writer, though, will ultimately call for an increased role in the decision-making processes of technological development. This case has already been made strongly for the user, both in this book and elsewhere (see Bravo 1993). Users are low on the proverbial totem pole and are rarely allowed access to the decision-making circle. Technical writers, I fear, are equally invisible in these circles where the design, development, and potential impact of technologies are (at least to some extent) determined. To employ writers in decision making is not that farfetched; it is at least not any more far-fetched than the aforementioned argument to rethink writers as interface designers. Consider the following scenario:

The Washington Metropolitan Transit Authority was experiencing a problem concerning the opening of their newly refurbished subway system. The system is a wonderfully designed, efficient technology ready to whisk people across the Washington metro area. An especially important aspect of the new system is the automated ticketing system. To use the system is, on a surface level, relatively easy. You choose your destination, locate the cost for the ride on a chart, deposit dollar bills in the automated ticket machine, and then you receive a ticket with the appropriate amount for your destination. A glitch, however, was noticed in the new automated ticketing system. Maybe "glitch" isn't the right word, because the problem wasn't in the mechanics of the system—this worked perfectly well. The problem was in the anticipated confusion that new users to the subway would have learning how to use the ticketing machines.

Imagine that you are one of several thousand commuters using the new system on your way to work. You have never used the ticketing machine, and as you stand in line waiting to purchase the ticket, you read the instructions posted beside the machine. As your turn to use the machine draws closer, you realize there are actually a number of tasks to carry out in order to be certain you have purchased the correct ticket for the correct price. Befuddled, you are jostled by others in the line and eventually lose your place in line. Finally, you figure out the procedure, but now you must wait for the next train.

During design evaluations of the system, the designers saw the possibility of this learning problem. How were they to be sure that thousands of new users of the system, in a hurry on their way to work, would use the new ticketing system (paid for, in part, through taxes levied on the new users)? The answer was ingenious. The transit authority hired a number of people to act as "experts" during the early days of operation of the new system. These "experts" were trained to use the ticketing machines, and when the system opened on its first day, they were there to act as expert users so the new users could observe. These "expert" ticket buyers merely went through the lines time after time so others could watch them. They were not in uniforms or any other official clothing: they merely looked and acted like users who knew what they were doing and who the "novices" could watch.

The decision to solve the problem of the ticketing machines in this way could easily have been from the minds of technical communicators: people who have an affinity for and are advocates of users. I am sure technical communicators had something to do with the writing of the instructions that sit beside the ticketing machines, but I am less sure that writers would have been involved with the decision to create "expert" users who serve as models. Such decisions are traditionally left to the designers, engineers, and developers: a tradition that portrays the developers as decision makers and writers as scribes. Once again, we see the issue of knowledge rearing its head. Who has it? How is it produced? In the final chapter we will conclude by taking this problem of knowledge into yet another context—the academic sphere. Here we will see the possibilites, and the inevitable limitations, that exist in the realm of the technical rhetorician.

CHAPTER 7

Technical Communication, Ethics, Curricula

User-Centered Studies and the Technical Rhetorician

> The important question about technology becomes, As we "make things work," what kind of *world* are we making?
>
> —Langdon Winner, *The Whale and the Reactor*

In the summer of 1992, a researcher working for the Physicians for Social Responsibility was gleaning documents recently released by the government pertaining to the early days of the American nuclear effort. There were thousands of pages, and the work was interesting in that it revealed story after story about how the American nuclear industry had evolved. Many of the stories, though, were not about the grand research sites we have come to know: the monstrous, secret factories and storage sites that served as nerve centers for the intricate nuclear complex—Hanford, Fernald, Rocky Flats. No, these stories were about hundreds of smaller sites, places where uranium was processed, machinery was milled, and materials were stored as they were moved across the country to the behemoths where the grand work of the nuclear industry was conducted.

One day this researcher came across a report that was shocking to him in a very personal way. The site—"The Alba-Craft Uranium Processing Site, Rose Avenue, Oxford, Ohio"—was at an address the researcher knew all too well: he had grown up just a few blocks away. (No one, of course, had ever talked about this site in his hometown, because the veil of secrecy that shrouded the entire nuclear movement in the 1950s was so opaque that even the workers at the sites often were unaware of the nature of their own work.) As he read on, he found that the site had been a very active uranium processing area; raw uranium had been milled into pellets and slugs from 1953 to 1958. During this short period, literally tons of uranium were processed, turned into "usable" uranium for the huge Fernald, Ohio, facility just a short jaunt down the railroad line.

The researcher's discovery lead him to call his mother, who happened to still live in Oxford, Ohio, and who was also the director of the local community action council.[1] Following the call, the director drove to Rose Avenue—a short, dead-end street abutting a main railroad trunk line—to discover that a warehouse existed on the site. The building was still used by a few local businesses to store merchandise, and one business, a screen printing company, used the facility for a workshop to make T-shirts.

The director of the center was alarmed, and maybe more so than most of us would be because she had for years (in her role as director of the Peace and Justice organization) been a student of the American nuclear buildup. She knew, in other words, that this apparently innocent warehouse was a potential area of radioactive contamination. How could it be anything but that, given what her son had told her? Tons of uranium? Dozens of workers? Transportation of radioactive materials? A virtually impenetrable veil of secrecy that had left the "truth" unknown for over three decades? The consequences were, to say the least, frightening.

The director, with the help of a number of concerned citizens and local government officials, took immediate action; over the next two years, the site was cleaned up through the concerted effort of many Oxford citizens. By the end of the project, the Department of Energy had leveled the building, removed contaminated soil, and essentially cleared the area of any radioactive materials. During this process, some technical communication students helped and, I contend, played the role of technical rhetoricians.

Discovering that uranium was processed down the street is scary stuff. But knowing that it happened is important—this is, of course, the intent of many "right to know" laws that have appeared in the last few years. From this knowledge, we—the citizens, users of our communities—can ask questions that hopefully will lead to more knowledge, and ultimately to as complete an understanding of an event as possible. The collection of this knowledge, however, is only a part of the process toward affecting change.

1. The name of the council is The Oxford Citizens for Peace and Justice, located on South Campus Avenue in the Campus Ministry Center.

In the case of uranium contamination in a residential district, affecting change means getting rid of the contaminants. Getting rid of the contaminants means gathering the information, putting it into appropriate forms, presenting it to numerous audiences, following up the presentation with more communications, "watchdogging" the cleanup itself, and, most important, reporting to the public all along the way so that the status of the project is made visible.

My role and the role of my students in this particular project was minimal. The real work of making sure the site was cleaned up was done by the director of the Oxford Citizens for Peace and Justice and some other community members. Nevertheless, a technical communication undergraduate class that I taught was involved in some of the early research stages that helped define, in part, the direction the cleanup might take. Through this class project, I witnessed students as they learned about the horrors of nuclear proliferation, the unwise use of natural resources, the blatant disrespect by previous governmental authorities for public safety in the 1950s, and the hazardous workplace conditions of workers in uranium processing sites during that early period of nuclear industry development.

This in itself is probably enough for a teacher to feel as though something good has happened in a class. In a technical communication class, however, it is not enough. The knowledge and awareness that students gain is certainly important, but *acting through* that knowledge is an equally important goal for technical communication pedagogy. Learning how to use communication media in the service of social action is central to nonacademic communication curricula. Further, it is the ability for students to learn how to act with a sense of responsibility, with an awareness of the *ethical* dimension of their actions that becomes a central issue in curriculum development. We face the same problem that many practical professions have faced as they emerge and develop the potential to exert more profound influence in social spheres. Thus, technical communicators are inevitably drawn to the same questions of ethics that other technical and scientific professions must ask: How can a practical profession—one that creates products that might alter human actions and possibly influence the way we live—ensure that it is acting responsibly? Is it even possible to know the potential consequences of our actions?

The complex maze of ethics has many historical pathways from which one can choose. One of these paths, the portions of the Aristotelian corpus devoted to ethics, is centrally fixed on the problem of action and thus serves as an excellent model for technical communicators. In the Aristotelian corpus, *acting* means taking social responsibility for one's actions—this is the essence of the Aristotelian notion of ethics. To know is one thing, but to act in the Aristotelian view means to do something that always has social consequences, or as it is explained in *The Nicomachean Ethics*, the end of action is in *the good*.

> Every art and every inquiry, and similarly every action and pursuit, is thought to aim at some good; and for this reason the good has rightly been declared to be that at which all things aim. (*Nicomachean Ethics*, 1094a1–4)

Developing curricula that incorporates "the good" is no easy task for the technical communication instructor. Much of our curriculum by necessity must teach the essential techniques of the profession—the arts of making that Aristotle refers to as the productive arts (see chapter 3). For instance, we are concerned that students learn the forms of technical documents, techniques for analyzing audience needs, the tools of visual design, and so on. Ethics and the problems of social action are difficult to pull into this already full classroom fray. Yet, we are constantly reminded by our colleagues Carolyn Miller, Paul Anderson, Steve Doheny–Farina, Marilyn Cooper, and Jim Porter, to name only a few, that the issue connecting technical communication to the social sphere most directly is ethics: the problem of analyzing the consequences of future actions.

As we saw in chapter 2, the end of rhetoric, and as I have argued the end of user-centered theory, is in the user. In this final chapter, I will argue that the end of user-centered theory is only complete when coupled with the end of social action. To paraphrase Aristotle, the end of the user-centered art, and similarly its every action and pursuit, is thought to aim at some good. This joining of productive knowledge and practical knowledge in the Aristotelian sense is the goal I wish to pursue. Developing curricula that meet this charge is no easy task, but the effort is well worth the attempt.

In the previous chapter, we surveyed in some detail a most likely place for the application of a user-centered theory: the context of computers and computer documentation. I say *most likely* because user-centered approaches arose as a result of computer industry design problems associated with the proliferation of mainframe computers and, especially, microcomputers. Consequently, it is appropriate that I have dedicated a significant amount of effort toward placing user-centered theory against the backdrop of computer technology. The central questions in this book, however, can be asked in many more contexts, and in many more modes, than through the analysis of interactions between users and the electronic medium. The study of technologies from a user perspective is tantamount to examining basic issues of how we as individual and social beings live: how we act with, against, and through technologies. In "Technologies as Forms of Life," Langdon Winner argues that these connections between technology and how we live have created a seamless phenomenon that is fruit for study in a variety of disciplines.

> Throughout their lives people come together to renew the fabric of relationships, transactions, and meanings that sustain their common existence. Indeed, if they do not engage in this continuing activity of material and social production, the human world would literally fall apart. . . . [F]rom this point of view, the important question about technology becomes, As we "make things work," [sic] what kind of world are we making? This suggests that we pay attention not only to the making of physical instruments and processes, although that certainly remains important, but also to the production of psychological, social, and political conditions as a part of any significant technical change. . . . Inquiries of this kind present an important challenge to all disciplines in the social sciences and humanities. Indeed there are many historians, anthropologists, psychologists, and artists whose work sheds light on long-overlooked dimensions of technology. Even engineers and other technical professionals have much to contribute here when they find courage to go beyond the narrow-gauge categories of their training. (*The Whale and the Reactor*, pp. 17–18)

The problem of attempting to understand the intersections of "psychological, social, and political conditions" and "significant technical changes" has been at the heart of this book, but I have reframed it in a way that could be more accurately termed "the problem of technological use": the problem that technology users (virtually all of us) face in their individual, social, historical, and political interactions with technology. To restate it as a problem of *use*, however, begs the presence of two distinct players in the technology game that Winner has neglected to mention: technical communicators and rhetoricians. This neglect, however, is not to be blamed on Winner. Instead, the blame sits squarely on our shoulders—those of us who participate in the loosely defined disciplines of technical communication and rhetoric. We, for the most part, have not made visible our presence, our potential influence in the sphere of technology studies.

There is considerable irony in this phenomenon. Technical communicators and rhetoricians are historically allied with action, and the fact that we are relatively slow to engage with the greater technology issues seems paradoxical. At the same time, however, it should not come as a surprise that those allied with the long historical lineage of everyday knowledge and mundane activity have been relatively uninfluential in regard to these issues. Technical communicators and rhetoricians are associated with devalued and marginal forms of productive and practical knowledge that have been relegated to the intellectual basement for over twenty-five centuries. My goal is to work with others in the profession of technical communication to point toward a more active role for technical communicators in the milieu of technology studies. To this end, I will focus on the sphere of academic curricula: a place where technical communicators and rhetoricians have some degree of influence.

In the remainder of this chapter, I will sketch a portrait of an academic sphere that encourages the development of a *technical rhetorician*: a technical communicator who is trained in the theory and practice of the arts of discourse, and who practices these arts as a responsible member of a greater social order. As the User-Centered Rhetorical Complex of Technology (see Figure 2.7) indicates, the greater social order circumscribes the entire complex of use in terms of cultural and historical situations. Thus in this final chapter we will explore the ramifications of these historical and cultural constraints on the teaching of technical communication. In the first section we will investigate a classroom environment that encourages direct participation with "real" audiences. In the second section I call for the development of technical communication curricula that is historically and theoretically informed but at the same time values practice.

User-Centered Pedagogy: Author(ity) and Ethics in the Classroom

About ten years ago I was a presenter on a conference panel focusing on the challenge of teaching technical communication in "real world" contexts. As I remember, all of the panelists discussed the classroom as a place where the "real world" could be simulated, to encourage student involvement with various situations of the workplace or other everyday communities. At the conclusion of the panel, during a question-and-answer period, a gentleman who was standing in the back of the room took issue (and rather sternly) with the panel's notion that the classroom itself is not the real world. He contended that the classroom *is* a real world: a place that is the everyday, real world of teachers and students. He claimed that the panel's concept of a real world set up an unfair binary, and as a result, we were cheapening the classroom because we relegated it to the weaker position.

In my thoughts I have returned to this moment several times over the intervening years, because it raises numerous issues for me about what it is that we—technical communication teachers—do in our everyday practice of teaching students to write and work effectively. On the one hand, I agree with the gentleman at the back of the room when he argues that we in fact are marginalizing the academic classroom as a kind of *other*. When the academic environment is refigured by technical communication instructors to reflect the world outside the classroom, we run the risk of making the classroom appear as something less than the nonacademic world. On the other hand, technical communication instructors are in a somewhat peculiar position when designing curricula because of technical communication's intimate connection with the nonacademic sphere. They must embrace the practices of the nonacademic world, at least to some degree, or else they could be labeled "charlatans."

The problem of accounting for the workplace reality of a student's post–school life is acute for technical communication instructors who teach inside many academic institutions. In these schools, the nonacademic environment can be represented as merely irrelevant to the educational mission of the academy,[2] or even more negatively as an environment that has highly questionable ethics. It is not uncommon, for instance, to have the concept of the workplace used as a site for teaching ethics, because it is assumed that the workplace is more likely to be a site of "the unethical." The academy and its classrooms are perceived as neutral safe havens, where "the unethical" can be fictionalized as something that occurs outside the walls of the classroom. Students are presented with a workplace situation that has fairly transparent ethical difficulties (for example, your boss has asked you to perform a task that covers up some crucial but possibly incriminating information), and they are asked to respond with a "solution" to the dilemma.

Such whistle-blower-type scenarios are not uncommon in technical communication pedagogy and often are found in *cases*—a phenomenon drawn from business school pedagogy. The case method presents students with narrative scenarios of problems similar to ones they might encounter in their post–school workplace lives. A case might ask for the solution to an ethical dilemma, but more often than not the student is asked to create a written document that solves a problem, or at the very least defines a problem so it can be acted upon at a later date. Although cases have declined in popularity to some extent in the past decade, they are still a strong "real world" tool of some programs' curricula.

More recently, technical communication pedagogists have adopted teaching strategies that aim to empower writers through a critical self-awareness of their beliefs vis a vis their situation as readers of texts. Advocates for this critical approach (which I will call the "critical readings approach") are centered in mainstream composition studies (Berlin 1988; Bizzell 1992; Trimbur 1989), but there is a growing number of those who argue for close readings of technical or public documents for the purpose

2. For an excellent discussion of the place of technical communication in the greater academy, see (Sullivan and Porter's 1993) "Remapping Curricular Geography."

of deconstructing the ideologically and culturally dominant drive of the text (see Wells 1986; Sullivan 1990).[3] A chief assumption of the critical readings approach is that students' consciousness will be expanded (or at least challenged) as they reflect upon the hierarchical authority represented in the text's diction, visual design, ideological assumptions, or other qualities inscribed in the text. Ideally, the critical readings approach purportedly takes on even more rigor when these deconstructive techniques are conducted in collaborative situations, for example, when peers respond to each others' interpretations and judgments during group interactions.

Learning to solve problems through cases or learning to critically analyze texts are certainly useful and important tactics for any technical communication student. Technical communicators, especially as they move into workplace or public situations, must be prepared to act quickly and prudently. I am unconvinced, however, that merely involving students with issues through texts is enough to engender a full enough understanding of how to *act* as technical rhetoricians. Simply put, to *know* does not necessarily translate to *know how.* I am reminded here of an old Bill Cosby stand-up routine that he performed in his early career. Cosby told the story of being on a college football team just before a big game, where the coach is addressing the players in the locker room. The coach gives an inspiring pep talk that "pumps up" the players to the point of near frenzy. When he finishes, however, the players rush to the locker room door—intent on "winning" and "overcoming all odds"—only to find that the door is locked and no one can find the key. By the time the key is found, the players have lost their inspiration and, it is assumed, they will lose the game as well.

I believe that in some crucial ways we run the risk of committing the same faux pas with our technical communication students if we rely too heavily on text-based models of learning, particularly if the desired end of our pedagogy is ultimately for the writers to take action: to move into the complicated realm of communication in the public and workplace spheres. Critical understandings of texts undoubtedly can engage students in social, political, and cultural problems and in turn will bring them to the metaphorical door, ready to take action. But will they have the "key," the rhetorical tools to attempt change through communication media? Or will they be limited to a self-aware knowledge that is lacking an ethic of action? For sure, we should not be interested in naively turning our students into the workplace only to "win" or "overcome the odds." The workplace/public

---

3. For a related discussion of the use of text-based approaches and technical communication, see the final chapter of Stephen Doheny–Farina's *Rhetoric, Innovation, Technology,* where he discusses the debate between Lipson and Parsons over the problem of teaching students about how to work across cultures.

sphere is too complex for such either/or solutions. Winning or losing are grossly incomplete metaphors for what often happens as people negotiate their way through problematic situations. At the same time, I think, we do not want our students to expand their consciousness and then become mere ballot box-filling or check-writing automotons who believe that the only effective ways to advocate for change are through the exercises of voting, or giving money to organizations who will take care of various social problems for them. The technical rhetorician would see tactics like voting and check writing, or of learning about social practices through the critical reading of texts, as being useful and necessary tools—part of any citizen's tool kit. But the technical rhetorician also will be aware of his or her power as a communicator who must become involved with actual people and situations: those mundane, everyday situations we are prone to overlook or take for granted.

That is, the technical rhetorician will be sensitive to user-centered concepts and how they shape the nature of public discourse. A user-centered approach to the classroom can help move from the awareness fostered by critical approaches toward active, socially involved pedagogy in at least three significant ways. First, a user-centered classroom is committed to the actual participation of the audiences (potential users) who will receive the communications being produced by the students. In the case of the Alba-Craft uranium facility mentioned in the opening scenario, the students met with original workers from the site, DOE officials, and many of the citizens involved with the cleanup operation. Through this interaction they obviously honed some research skills as they interviewed these participants and scoured reams of technical government documents, but they also had to negotiate with the audiences and take their views seriously. The documents the students were creating would be read, and eventually used, by these audiences. The pressure was on to be accountable, to be ethical in a personal sense. More than just a grade in the class was at stake.

While not every project in a given class can attain such goals, it is important that some projects do so because the outcome of such projects allows students to see, for possibly the first time in their educational careers, how they can have a direct effect on the social sphere through communication production. Certainly there is a danger in leading students down a path of naive hope and optimism. As researchers, including Nancy Roundy Blyler, Dale Sullivan, and Craig Waddell have aptly pointed out, the probability of everyday citizens having power in the public or workplace sphere is low. These views, however, are tinged with a pessimistic determinism that might send instructors to the classroom believing that text-based models are the only refuge in this technologically controlled society.

From a user-centered perspective, I argue, we do have a choice whether to act according to our beliefs, especially when these beliefs are set in juxtaposition to technological "progress." To rephrase Langdon Winner's statement earlier in this chapter, if we do *not* engage in this continuing activity of material and social production, the human world could literally fall apart. To *accept* the determinist logic that what will happen is inevitable is, in a word, static; to *confront* the logic of determinism is dynamic: a movement (as small as it may appear) toward the "two-way harmony" of Habermasian theory so often cited by technical communication scholars searching for answers to the problems of communicative action.[4]

Second, the user-centered approach advocates the *direct* involvement of the audience in the discourse production process. Ironically, the physical presence of the intended audience may actually diffuse some of the authority that the writers have over the discourse being produced. This alteration of power might be most difficult for the writing instructor to swallow. Indeed, it is probably just as difficult for a writing teacher to accept the loss of authority by writers as it is for systems developers to accept loss of authority to end users. The argument, however, is virtually the same. Writers of technical and scientific documents, particularly those documents that have direct impact on users, must be compelled to share the authority of text equally with the user audience. Most current approaches to the writing classroom will just not allow for this shift in power. For instance, in a critical readings approach, the audience is most often represented through text (print or electronic). Consequently, writers in these situations retreat to traditional conceptions of writing that promote the writer as the author(ity). The user, the end of discourse from a rhetorical point of view, becomes a fiction who is imagined in terms of the writer's mental construction.

I would be foolish to claim that the "fiction" of the audience would ever disappear from the writer's perspective: that is far from my intent.[5] Instead I claim that the engagement of writers with real audiences and actual[6] users creates an environment for text production (here I use the word "text" broadly) that alters the authority of teachers and writers alike—

---

4. This is a reference to Nancy Roundy Blyler (1994) and Susan Wells (1986.)

5. For more on the concept of a fictional audience, see Ong 1975 or Ede and Lunsford 1984.

6. I want to be clear that by "actual" I mean that students will interact with audiences directly. This interaction is not limited to face-to-face contact, but can be through electronic or other media. The central issue for me here is one of negotiation between writers and their audiences, regardless of the medium.

an alteration that forces a refiguring of the ends of discourse toward the users. The text is generated through an actual interaction between the user and the writer. Agency, in other words, is not lost in the user-centered paradigm, it is openly shared. As I have already said, this actual interaction cannot occur in every classroom project, nor will it occur with great frequency in many of the students' professional lives. There will be many times when these writers will find themselves behind the walls of an office cubicle writing to an audience they barely know. This also raises the problem of convincing industry that actual interaction with audiences should be more aggressively promoted. Opening the classroom to real audiences certainly provides students with experiences they might encounter in the workplace. Industry, however, is not always ready to promote direct access to potential audiences; the fictional audience is just more cost-effective. The short-term profit motive, as I discussed in the previous chapter, stifles industry practice in terms of promoting actual audience/technical communicator interaction. Students should not be led to believe that they will always have the luxury of actual audience interaction. They might, however, become informed advocates for revision of industry practice from the inside, similar to arguments that have been made regarding the user advocacy role of usability specialists.

The ill-defined[7] nature of actual situations may also alter the often held student conception that they are merely writing for the teacher. Teachers in the user-centered classroom will find that they are able to step back from the role of being the central audience. Teachers often complain that students only write "what the teacher wants." I have found repeatedly that actual audiences and actual situations deflect the attention given to the teacher, thus freeing the teacher to become an active player in the production process as opposed to being only an objective evaluator of the students' processes and texts.

Third, and maybe most important, students might learn that solving communication problems is often a matter of time: doing a little at a time instead of looking for the final solution prematurely or all at once. Case-inspired pedagogy is most guilty of setting up expectations that communication problem solving is a matter of finding the right answer, or being able to solve the whole problem. Students also are often led by case pedagogy to create certain formats of documents for certain problems. While there is a logic to this practice when introducing document types to students, there is an inherent danger in that students expect the formats themselves to fill

---

7. For an excellent discussion of ill-defined problem solving in educational settings, see chapter 1 of *The Complete Problem Solver* by John Hayes.

audience needs. The genre, in other words, appears to solve the communication problem, and the student falsely perceives the communication production effort to be a format-filling exercise. The students involved with the Alba-Craft cleanup, for example, experienced a situation where genres were not readily apparent. They could not just write a specific type of report for a particular audience because the problem was so ill-defined and the audiences often were overlapping or just plain hard to identify. To determine the communication needs, they surveyed the breadth of the problem, honed in on important facets of the problem, and then designed communications that would enable the audience to move forward. In some cases this meant writing explanations of policy for publication in newspapers; in other cases it called for interviewing former employees of Alba-Craft and then reporting on those findings for members of the Peace and Justice organization; in yet other scenarios the students charged themselves with writing persuasive documents to aid in implementing the timetable of the cleanup. They also often exchanged research with one another and learned how to adapt the same information for different audiences and purposes.

The instructional practice of dictating format or genre types as solutions to communication problems is probably the basis for those widespread criticisms often waged against technical communication instruction—that all we do is have students fill in prescribed formats in uncreative, unreflective ways. These criticisms of technical communication pedagogy are unfortunate, but we may beg for the abuse of such criticisms if we continue to foster pedagogical exercises that appear to dictate formats. We do not, of course, want to throw the baby out with the bath water. By that I mean the teaching of formats and document structures is an important part of technical communication instruction. Many genres and formats have been developed over the years, in part, due to audience research (see Dodge 1962; Redish 1988), and we should resist efforts to dismiss document genres as merely controlling or uncreative boilerplate. We should instead synthesize genre knowledge with social knowledge to create situated pedagogy: pedagogy that is conscious of its contingent and mutable nature.

History and Theory in Technical Communication Curricula

The history of technical communication is for the most part still waiting to be written (Moran 1985; Tebeaux and Killingsworth 1993) and, as a consequence, few disciplines tell so little history of their field to their students as we do in technical and scientific communication. Technical communicators are not the first professionals to suffer from a lack of disciplinary history. In the 1950s, engineering programs were concerned about similar issues of their history. To meet this shortfall, engineering programs integrated some history of engineering into the curricula, but these attempts were not without problems, as many schools merely enacted service courses that glossed over the history of the profession. We should be wary and take a lesson from the engineering profession, both from their successes and failures in attempting to place history into their curriculum. History, after all, is nearly always interesting and for that reason alone technical communicators can justify pursuing historical pedagogy. But, as I will explain shortly, history does not always need to be presented as some grand affair: a story limited to the great accomplishments of great people at opportune moments. History is in the everyday, mundane activity of daily experience the technical communicator deals with all the time. We can bring history to the classroom in many forms.

The same can be said of theory in the technical communication classroom—it has many forms, some with an upper-case "T" and some with a lower-case "t." Technical communication has been defined as a practical discipline, and rightly so. Technical communicators practice their craft daily, and as we have seen in the previous discussion of "real" contexts, we carry out these actions in the public sphere. Practice in the profession of technical communication is valued highly because we understand that practice has strong ethical and political dimensions; it is not just mere doing, but it is doing that has social consequences. Because we value practice so highly, however, does not also mean that we should hold theory at arm's length—like some sort of unwanted, foul-smelling thing that has been thrown in our direction. Theory is everywhere. It evolves from practice, as I would hope that this book itself has made clear. We should define it and use it, and we should do this in our own way. That is, as we borrow theory from other places, or discover those theories that we might actually be able to call "our own," we should readily admit that theories are useful to the development of our understandings of what it is we do as a "practical" profession.

I will make the case that scenarios such as the one presented next are replete with theoretical and historical matters that directly affect the practices of students in technical and scientific communication programs, and eventually their professional life. I contend that students can use such tales to explore systematic knowledge in scientific or technological endeavor, to critique the problems of constructing "reality" through scientific "facts," to examine the role of political, economic, or ideological factors in techniques, and to apply what they learn to their everyday practices as technical or scientific communicators. In short, I suggest that students who are grounded in theoretical and historical concepts of technical communication will be better able to work as ethical, active, and fulfilled members of public and/or private institutions.

In the heat wave of July 1993, a City of Philadelphia health examiner was called to the low-income, high-rise apartment of an elderly woman who had been found dead. The body had already gone into some early stages of rigor mortis, thus such information as the body temperature at the time of death was impossible to ascertain. The examiners problem: What was the cause of death?

To determine the cause of death, the examiner was required to follow a procedural technique, then make a good guess based upon medical "facts." The technique called for the examiner to use only the information that was verifiable—therefore, a guess at body temperature was not allowed because the person had been dead too long. Thus, in many instances, if the examiner used the required technique, the determined cause of death was likely to be something other than excessive heat because only those "facts" that were "true" or verifiable were admissible as evidence in determining the cause of death.

In 1993, however, the chief examiner of Philadelphia decided to "bend the rules" and use information that was not usually part of the normal procedure. For example, he asked his examiners to "act like detectives" and collect information that went beyond the prescribed procedure. For instance, they found similarities in a number of the cases.

- *the outside temperature had been well in excess of 90º Fahrenheit*
- *the apartment temperature was also high—about 85º*
- *there was an air conditioner in the room but it had not been turned on*
- *the windows had been sealed shut, possibly to deter burglars*
- *the person lived alone, and was found because a manager checked*

As a result, the City of Philadelphia reported a much greater instance of death from heat than any other major city in the United States (118 during a ten-day period, while New York and Washington, D.C., had about ten each). The Philadelphia examiner was heavily criticized for the increased numbers and, in particular, he was accused of violating accepted scientific methods of determining "cause of death." However, after a year of investigation by the Disease Control Center in Atlanta, it was concluded that indeed the Philadelphia examiner was justified (or at least not "wrong") in his new procedures. This tale, of course, has not totally been told yet. The old procedure must be reviewed, there will be much debate, and maybe something will change (already the Philadelphia authorities are making fans more accessible and are discussing how to revise the definition of "heat emergency").

What could technical communication students learn from such a "tale"? How might the use of stories like this one enrich what we do in the classroom—both practically and theoretically? To begin, students could be introduced to the concepts of social constructionist theories and methods. It is clear from the Philadelphia tale that the definition of death, or at least the cause of death, is mutable. What was a cause of death for one city government was not for another, thus the definition of death was shaped by context. For instance, the result of the Philadelphia definition took on greater consequences and eventually broader action (such as making fans more accessible).

The use of historical examples can enliven the technical writing class in ways that I do not think we are prone to do now. Wedded with user-centered classroom projects like those I have described earlier, readings into historical issues can bring the best aspects of the critical readings approach forward. Although the Philadelphia scenario is not an explicit example of the history of technical communication, it is nevertheless a story that raises historical issues that should be closely examined by technical communication students. For instance, the issues of "facts," "objective knowledge," and "truth" are easily spotted in this scenario. The medical examiner's redefinition of "death" flies in the face of traditional definitions of scientific fact. If death cannot even be defined with some certainty, then what can? Classes engaged in this issue alone could easily find themselves exploring the Royal Society of the seventeenth century and their obsession with transparent language and scientific objectivity. They also could investigate the nature of the scientific method itself and possibly compare that method to some common to technical communicators, such as those associated with problem solving.

In technical and scientific communication curricula, we have a unique opportunity to engage students in theoretical and historical concepts as we teach students who are *making the artifacts of their profession*: they are actually creating the products of their "trade." They are not just learning about the artifacts or practices, but they are actually creating the products that exemplify theory. Consequently, students can engage firsthand in the theoretical concepts and historical scenarios as they create their communication products. They can draw analogies to the actions of other technicians or scientists—such as the medical examiners who were told to alter their strict scientific practices and, instead, "act like detectives." Technical writers working in usability situations, for instance, often act like detectives, but I am not sure that they often think of their actions in this way. Instead, like the traditional medical examiners, they probably act in less flexible, more rule-bound procedural ways.

In our attempt to incorporate historical and theoretical elements, we should not ignore the immediate history of our profession as a way to demonstrate the various ideological threads that make up the technical communication tapestry. We need to look, for instance, no further than our own professional organizations for such material. A prime example is the decade-old debate between the Society for Technical Communication (STC) and the Council of Programs in Technical and Scientific Communication (CPTSC) over the issue of academic program accreditation. The STC, serving as the voice of industry interests, has called for an accreditation process for programs in technical communication—something akin to the accreditation of engineering and other professional or technical programs. They see accreditation as a way to ensure that writers coming from academic programs are trained to meet industry needs. The CPTSC, largely representing the academic program constituency, argues that accreditation would severely limit programs' abilities to push the proverbial envelope of technical communication practices: to test theoretically sound, yet unproven, techniques. As an alternative, the CPTSC has suggested that programs go through an evaluation process that would openly discuss programs' defining features, but would not ultimately lead to a simple accreditation rating. The majority of CPTSC members argue that evaluations could provide industries with substantial information upon which they could make their own judgments pertaining to hiring different programs' graduates.

This is not an easily resolvable issue and would be fodder for interesting classroom debates that might urge student exploration of the publications of professional organizations, thus bringing students to publications for more than just the "how-to" information they might contain. The ideological biases of such organizations can be discussed, too, and serve as an important site for the basic differences in our field. To dispel the myth of disciplinary monoliths, regardless of the discipline, is always a healthy endeavor.

Most significantly, the integration of technical communication history and theory into curricula will eventually create more reflective students: students who can make ethical and practical judgments with a well-formed base of knowledge about what it means to create and disseminate technical and scientific communication. In the case of the Philadelphia Health Department, I believe technical communicators can see themselves rewriting policy and procedures that actually will have an effect on the public sphere (for instance, compelling authorities to provide more fans for the populace that needs them). Students might also be able to see ethical problems as being more than two-pronged dilemmas that compel an individual to always carry the burden of "whistle-blower." Instead, they might see their role as being communal actors in a larger scenario of actions that can, at times, offer alternatives to the "whistle-blower/not whistle-blower" dichotomy we so often promote in our treatment of ethics.

The aforementioned discussion has been primarily concerned with technical communication pedagogy as it relates to "practical" curricula in both undergraduate and graduate (master's) programs. This has been the charge of most technical communication programs and will likely remain their focus for years to come. With the growth of these programs has also come the need for faculty to teach in them, which has, in turn, opened the door for graduate study that prepares technical communicators as research scholars: academic professionals whose work is more or less evenly divided between teaching and research and who must conduct research in order to be deemed tenurable by research-oriented institutions. With this demand for researchers has come a growth in graduate programs at the doctoral level. While my purpose is not to pursue the nature of these programs in depth here, I do want to point out that the inculcation of historical and theoretical subject matter will be even more pressing as time goes on. The development of curricula that walks a fine line between theory and practice, or that succeeds in refiguring the line, will take on new meaning in institutions that must meet these emerging needs. Indeed, a remaking of the theory/practice binary within educational and nonacademic spheres may be the ultimate reward for embracing the thorny notions of user-centerdness. It may be in fact, such fundamental challenges to modern belief systems that will provide a potential voice for users within the greater arena of technology studies.

References

Adams, Jack A. 1965. *Human Factors Engineering.* New York: Macmillan Publishing.

Adler, Paul, and Terry A. Winograd. 1992. *Usability: Turning Technology into Tools.* New York: Oxford University Press.

Aeschylus. 1966. *Prometheus Bound.* Translated by George Thomson. Cambridge: Cambridge University Press.

American Institutes of Research. 1982. "The Process Model of Document Design." In *Simply Stated.* Washington, D.C.

Anderson, J. R. 1983. *The Architecture of Cognition.* Cambridge, Mass.: Harvard University Press.

Anderson, Paul. 1994. *Technical Writing: A Reader-Centered Approach.* New York: Harcourt Brace Jovanovich.

Aristotle. 1952. *The Works of Aristotle, Vol. 1 & 2.* R.M. Hutchins, ed. Chicago: *Encyclopedia Brittannica,* The University of Chicago Press.

Atwill, Janet. 1993. "Instituting the Art of Rhetoric: Theory, Practice and Productive Knowledge in Interpretations of Aristotle's *Rhetoric.*" In Takis Poulakis, ed. *Rethinking the History of Rhetoric: Multidisciplinary Essays on the Rhetorical Tradition.* Boulder, Colo.: Westview Press.

Baecker, R., and W. Buxton, eds. 1987. *Readings in Human–Computer Interaction: A Multidisciplinary Approach.* Los Altos: Morgan Kaufmann Publishers.

Bannon, L. J. 1986. "Issues in Design: Some Notes." In D.A. Norman, and S.W. Draper, eds. *User-Centered System Design: New Perspectives on Human–Computer Interaction* (pp. 25–29). Hillsdale, N.J.: Lawrence Erlbaum Assoc., Publishers.

Barker, Thomas T. 1998. *Writing software documentation: A task-oriented approach.* Needham Heights, MA: Allyn & Bacon.

Barrett, Edward. 1989. *The Society of Text: Hypertext, Hypermedia, and the Social Construction of Information.* Cambridge, Mass.: MIT Press.

Barton, Ben F., and Marthalee S. Barton. 1989. "Trends in Visual Representation." In Charles Sides, ed. *Technical and Business Communication: Bibliographic Essays for Teachers and Corporate Trainers* (pp. 95–135). Urbana, Ill.: NCTE.

Bazerman, Charles. 1988. *Shaping Written Knowledge: Essays in the Growth, Form, Function, and Implications of the Scientific Article.* Madison, Wis.: Wisconsin University Press.

Bazerman, Charles, and James Paradis. 1991. *Textual Dynamics of the Professions: Historical and Contemporary Studies of Writing in Professional Communities.* Madison, Wis.: Wisconsin University Press.

Bell, Paula, and Charlotte Evans. 1989. *Mastering Documentation.* New York: John Wiley & Sons.

Beniger, James R. 1986. *The Control Revolution: Technological and Economic Origins of the Information Society.* Cambridge, Mass.: Harvard University Press.

Berlin, J. 1988. "Rhetoric and Ideology in the Writing Class." *College English* 50: 477–94.

Bernhardt, Stephen A. 1986. "Seeing the Text." *College English* 37(1): 66–78.

———. 1993. "The Shape of Text to Come: The Texture of Print on Screens." *College Composition and Communication* 44(2): 151–75.

Berry, Wendell. 1973. *The Unsettling of America.* Los Angeles: Sierra Club Books.

Bijker, Wiebe. 1993. "Do Not Despair: There Is Life After Constructivism." *Technology, Science, and Human Values* 18(1): 113–38.

Bijker, Wiebe E., Trevor P. Hughes, and Trevor Pinch. 1987. *The Social Construction of Technological Systems: New Directions in the Sociology and History of Technology.* Cambridge, Mass.: MIT Press.

Bijker, Wiebe, and John Law. 1992. *Shaping Technology/Building Society: Studies in Sociotechnical Change.* Cambridge, Mass.: MIT Press.

Bizzell, Patricia. 1992. *Academic Discourse and Critical Consciousness.* Pittsburgh: University of Pittsburgh Press.

Bizzell, Patricia, and Bruce Herzberg. 1990. *The Rhetorical Tradition: Readings from Classical Times to the Present.* New York: St. Martin's Press.

Blyler, Nancy R. 1994. "Habermas, Empowerment, and Professional Discourse." *Technical Communication Quarterly* 3(2): 125–46.

Boden, M.A. 1988. *Computer Models of Mind: Computational Approaches in Theoretical Psychology.* Cambridge: Cambridge University Press.

Bødker, Susanne. 1989. "Through the Interace: A Human Activity Design Approach to Computer Interface Design." *Human Computer Interaction* 4(3): pp 171–95.

Bolter, Jay D. 1991. *Writing Space: The Computer, Hypertext, and the History of Writing.* Hillsdale, N.J.: Lawrence Erlbaum.

Bourdieu, Pierre. 1977. *An Outline of a Theory of Practice.* Translated by Richard Nice. Cambridge: Cambridge University Press.

Bradford, Annette N. 1989. "A Planning Process for On-line Information." In Doheny–Farina, ed. *Effective Documentation: What We Have Learned From Research* (pp. 183–211). Cambridge, Mass.: MIT Press.

———. 1984. "Conceptual Differences Between the Display Screen and the Printed Page." *Technical Communication* 31: 13–16.

Bravo, Ellen. 1993. "The Hazards of Leaving Out the Users." In Douglas Schuler and Aki Namioka, ed. *Participatory Design: Principles and Practices* (pp. 3–11). Hillsdale, NJ: Lawrence Erlbaum.

Britton, J. 1970. *Language and Learning.* London: Penguin Books.

Brockman, John R. 1991. *Writing Better Computer User Documentation.* New York: John Wiley & Sons.

———. 1988. "Desktop Publishing—beyond GEE WHIZ." *IEEE Transactions on Professional Communications* 31(1): 21–29.

Brown, J., and S. Newman. 1985. "Issues in Cognitive and Social Ergonomics: From Our House to Bauhaus." *Human–Computer Interaction,* vol. 1: pp. 359–91.

Brown, J., A. Collins, and P. Duguid. 1988. *Situated Cognition and the Culture of Learning.* Report No. IRL 88-0008. Palo Alto, Calif.: Institute for Research on Learning.

Brown, John S., and Paul Duguid. 1994. "Borderline Issues: Social and Material Aspects of Design." Human–Computer Interaction 9(1): 3–36.

Bunge, Mario. 1985. "Technology: From Engineering to Decision Theory." In *Treatise on Basic Philosophy,* vol. 7. *Epistemology and Methodology III: Philosophy of Science and Technology,* part 2, "Life Science, Social Science, and Technology." Boston: D. Reidel, pp. 219–311.

Burgess, J. H. 1989. *Designing for Humans: The Human Factor in Engineering.* Princeton, N.J.: Petrocelli Books.

Burnet, John. 1964. *Greek Philosophy: Thales to Plato.* New York: St. Martin's Press.

Bush, V. 1945. "As We May Think." *Atlantic Monthly* (August): 101–08.

Carroll, J. 1987a. "Minimalist Design for Active Users." In R. Baecker and W. Buxton, eds. *Readings in Human–Computer Interaction: A Multidisciplinary Approach* (pp. 621–26). Los Altos, Calif.: Morgan Kaufman Publishers.

———. 1987b. "The Adventure of Getting to Know a Computer." In R. Baecker, and W. Buxton, eds. *Readings in Human–Computer*

Interaction: A Multidisciplinary Approach (639–48). Los Altos, Calif.: Morgan Kaufman Publishers.

———. 1990. *The Nurnberg Funnel: Designing Minimalist Instruction for Practical Computer Skill.* Cambridge, Mass.: MIT Press.

Carroll, John and Robert L. Campbell, 1986. "Softening up Hard Science: Reply to Newell and Card." *Human–Computer Interaction* 2: 227–49.

Carroll, John and M.B. Rosson. 1987a. "The Paradox of the Active User." In J.M. Carroll, ed. *Interfacing Thought: Cognitive Aspects of Human–Computer Interaction* (pp. 80–111). Cambridge, Mass.: MIT Press/Bradford Books.

Chapanis, A. 1965. "Words, Words, Words." *Human Factors* 7: 1–17.

———. 1979. *Man–Machine Engineering.* London: Tavistock Publications, Ltd.

Charney, D., and L. Reder. 1986. "Designing Interactive Tutorials for Computer Users." *Human–Computer Interaction,* vol. 2: 297–317.

Charney, D., L. Reder, and G. Wells. 1988. "Studies of Elaboration in Instructional Texts." In S. Doheny–Farina, ed. *Effective Documentation: What We Have Learned From Research* (pp. 47–71). Cambridge, Mass.: MIT Press.

Christensen, J. M. 1958. "Trends in Human Factors." *Human Factors:* 1: 2–7.

Cicero. 1942. *De Oratore.* Trans. E. W. Sutton. Cambridge, Mass: Harvard University Press.

Cockburn, Cynthia. 1985. *Machinery of Dominance: Women, Men and Technical Know- how.* London: Pluto Press.

Crandall, J.A. 1987. *How to Write Tutorial Documentation.* Englewood Cliffs, N.J.: Prentice Hall, Inc.

Curtis, M.S. 1988. Windows on Composing: Teaching Revision on Word Processors. *College Composition and Communication* 39: 337–34.

D'Angelo, F. J. 1975. *A Conceptual Theory of Rhetoric.* Cambridge, Mass.: Winthrop Publishers.

de Certeau, Michel. 1984. *The Practice of Everyday Life.* Berkeley, Calif.: California University Press.

Detienne, Marcel, and Jean-Pierre Vernant. 1974. *Cunning Intelligence in Greek Culture and Society.* Translated by Janet Lloyd. Chicago: The University of Chicago Press.

Dieli, M. 1988a. "How Can Technical Writers Effectively Revise Functional Documents?" In Beene and P. White, eds. *Solving Problems in Technical Writing* (pp. 150–66). New York: Oxford University Press.

———. 1988b. "A problem solving approach to usability test planning." *ITCC Proceedings.* Washington, D.C.: Society for Technical Communication, pp. 89–91.

Dobrin, D. 1983. "What's Technical About Technical Writing?" In P. Anderson, R. Brockman, and C. Miller, eds. *New Essays in Scientific and Technical Writing* (pp. 227–50). Farmingdale, N.Y.: Baywood.

Dodge, Richard W. 1962. "What to Report." *Westinghouse Engineer,* July Sept.: pp. 3–7.

Doheny–Farina, Stephen. 1984. "Writing in an Emergent Business Organization: An Ethnographic Study. (Doctoral Dissertation, Rensselaer Polytechnic Institute.) *Dissertation Abstracts International* 45, 3337A.

———. 1992. *Rhetoric, Innovation, Technology: Case Studies of Technical Communicators in Technology Transfers.* Cambridge, Mass.: MIT Press.

———. 1996. *The Wired Neighborhood.* New Haven: Yale University Press.

Draper, S.W. 1985. "The Nature of Expertise in UNIX." In R. Baecker and W. Buxton, eds. *Readings in Human–Computer Interaction: A Multidisciplinary Approach* (pp. 621–26). Los Altos, Calif.: Morgan Kaufman Publishers.

Dreyfus, Hubert L. 1972. *What Computers Can't Do.* New York: Harper & Row.

Dreyfus, Hubert, and Stuart Dreyfus, with Tom Athanasiou. 1986. *Mind Over Machine: The Power of Human Intuition and Expertise in the Era of the Computer.* New York: Free Press.

Duffy, T. M. 1985. "Readability Formulas: What's the Use?" In T.M. Duffy and R. Waller, eds. *Designing Usable Text* (pp. 113–43). Orlando, Fla.: Academic Press.

Duffy, T. M. and R. Waller, eds. *Designing Usable Text.* Orlando, Fla.: Academic Press.

Duffy T. M., B. Mehlenbacher, and J. Palmer. 1989. "The Evaluation of Online Help Systems: A Conceptual Model." In E. Barrett, ed. *The Society of Text: Hypertext, Hypermedia, and the Social Construction of Information* (pp. 362–87). Cambridge, Mass.: MIT Press.

Duin, A. H. 1989. "Factors That Influence How Readers Learn From Text: Guidelines for Structuring Technical Documents." *Technical Communication* 36, 2: 97–101.

———. 1993. "Test drive—Evaluating the Usability of Documents." In Carol Barnum and Saul Carliner, eds. *Techniques for Technical Communicators* (pp. 306–27). Macmillan Publishing Co.

Dumas, Joseph S., and Janice C. Redish. 1994. *A Practical Guide to Usability Testing.* Norwood, N.J.: Ablex Publishers.

Dunne, Joseph. 1993. *Back to the Rough Ground: 'Phronesis' and 'Techne' in Modern Philosophy and in Aristotle.* Notre Dame, Ind: Notre Dame University Press.

Ede, Lisa, and Andrea Lunsford. 1984. "Audience Addressed/Audience Invoked: The Role of Audience in Composition Theory and Pedagogy. *CCC* 35: 155–71.

Ehn, Pelle. 1988. *Work-Oriented Design of Computer Artifacts.* Stockholm: Arbetslivscebtrum.

Elbow, Peter. 1987. "Closing My Eyes as I Speak: An Argument For Ignoring Audience." *College English* 37: 294–301.

Ellul, Jacques. 1964. *The Technological Society.* New York: Vintage Books.

Fawcett, H. 1989. "Using Tagged Text to Support On-line Views." In *ITCC Proceedings* (rt79–81). Society for Technical Communication: Washington, D.C.

Feenberg, Andrew. 1991. *Critical Theory of Technology.* Oxford: Oxford University Press.

Felker, D. B., ed. 1981. *Document Design: A Review of the Research.* Washington, D.C.: American Institutes of Research.

Fitzgerald, Deborah. 1993. "Farmer's Deskilled: Hybrid Corn and Farmer's Work." *Technology and Culture* 34(2): 324–43.

Flower, Linda. 1993. *Problem-Solving Strategies For Writing.* 24th ed. New York: Harcourt Brace Jovanovich.

———. 1989. "Cognition, Context and Theory Building." *College Composition and Communication* 40(3): 282–311.

———. 1993. "Cognitive Rhetoric: Inquiry Into the Art of Inquiry." In Theresa Enos and Stuart C. Brown, eds. *Defining the New Rhetorics* (pp. 171–90). Newbury Park: Sage.

Floyd, C., et al. 1989. "Out of Scandinavia: Alternative Approaches to Software Design and Development. *Human–Computer Interaction* 4: 253–349.

Forester, T. 1989. *Computing in the Human Context: Information, Productivity and People.* Cambridge, Mass.: MIT Press.

Foucault, M. 1973. *The Order of Things: An Archaeology of the Human Sciences.* New York: Vintage Books.

Fuller, Steve. 1993. *Philosophy of Science and Its Discontents.* New York: The Guilford Press.

Gardiner, M., and B. Christie. 1987. *Applying Cognitive Psychology to User–Interface Design.* Chichester: John Wiley & Sons.

Geertz, Clifford. 1983. *Local Knowledge: Further Essays in Interpretive Anthropology.* New York: Basic Books.

Gore, Al. 1994. "Democracy and Technology." *Discover: The World of Science* (October): 38–46.

Grice, Roger. 1989a. "Interactive Interfaces: Dimensions and Measurable Characteristics." In *ITCC Proceedings.* Society for Technical Communication, rt63–66.

———. 1989b. "Document Development in Industry." In B.E. Fearing and W.K. Sparrow, eds. *Technical Writing: Theory and Practice* (pp. 27–32). New York: Modern Language Association.

Grice, R. A. 1987. "Technical Communication in the Computer Industry: An Information–Development Process to Track, Measure, and Ensure Quality." *Dissertation Abstracts International* 49, DA8803441.

Grimm, S. J. 1982. "*How to Write Computer Manuals for Users.*" Belmont, Calif.: Lifetime Learning Publications.

Gurak, Laura J. 1997. *Persuasion and Privacy in Cyberspace: The On-line Protests Over Lotus MarketPlace and Clipper Chip.* New Haven: Yale University Press.

Halloran, S. Michael. 1975. "On the End of Rhetoric, Classical and Modern." *College English* 36: 621–31.

Haraway, Donna. 1991. *Simians, Cyborgs and Women: The Reinvention of Nature.* New York: Routeldge.

Hartley, James. 1985. *Designing Instructional Text* 2d ed. London: Kogan Page Ltd.

Haselkorn, M. P. 1988. "The Future of 'Writing' for the Computer Industry. In E. Barrett, ed. *Text, Context, Hypertext: Writing With and For the Computer* (pp. 3–13). Cambridge, Mass.: MIT Press.

Hastings, G. P., and K. J. King. 1986. *Creating Effective Documentation for Computer Programs.* Englewood Cliffs, N.J.: Prentice Hall, Inc.

Hayes, John R. 1981. *The Complete Problem Solver.* Philadelphia: Franklin Institute.

Heidegger, Martin. 1962. *The Question of Technology and Other Essays.* Translated by William Lovitt. San Francisco: Harper and Row.

Herbig, P. A. and H. Kramer. 1992. "The Phenomenon of Innovation Overload." *Technology in Society,* vol. 14: 441–61.

Hightower, James. 1978. *Hard Tomatoes, Hard Times.* Cambridge: Schenkman.

Horton, W. 1988. "Myths of On-line Documentation." In *ITCC Proceedings* (ata 43–46). Society for Technical Communication: Washington, D.C.

Horton, W. 1990. *Designing and Writing On-line Documentation: Help Files to Hypertext.* New York: John Wiley & Sons.

Johnson, Bob. 1990. "User-Centeredness, Situatedness, and Designing the Media of Computer Documentation." *Proceedings of SIGDOC '90* 14(4): 55–62.

Johnson, R. R. (March 1989). "The 'Art' of Visual Rhetoric." Paper presented at the Conference on College Composition and Communication, Seattle, Wash.

———. 1994. "The Unfortunate Human Factor: A Selective History of Human Factors for Technical Communicators. *Technical Communication Quarterly* 3(2): 195–212.

———. 1995. "Romancing the Hypertext." *Technical Communication Quarterly* 4(1): 11–22.

———. 1997. "Complicating Technology: Interdisciplinarity, The Burden of Comprehension, and the Ethical Space of the Technical Communicator." *Technical Communication Quarterly* 7(1): 1–24.

Johnson, Robert. 1935. King of the Delta Blues: The Complete Recordings. New York: CBS Records.

Johnson, H. Thomas, and Robert S. Kaplan. 1987. *Relevance Lost: The Rise and Fall of Management Accounting.* Boston, Mass.: Harvard Business School Press.

Jonassen, D. 1985. *The Technology of Text: Principles for Structuring Designing and Displaying Text* vols. 1 & 2. Englewood Cliffs, N.J.: Educational Technology Publications.

Jones, J.C. 1967. "The Designing of Man–Machine Systems." *Ergonomics* 10(2): 101–11.

Jones, M.K. 1989. *Human–Computer Interaction: A Design Guide.* Englewood Cliffs, N.J.: Educational Technology Publications.

Kalmbach, J. 1988. "Technical Writing Teachers and the Challenges of Desktop Publishing." *The Technical Writing Teacher* XV, 2: 117–31.

Katz, Steven B. 1993. "Aristotle's Rhetoric, Hitler's Program, and the Ideological Problem of Praxis, Power, and Professional Discourse." *Journal of Business and Technical Communication* 7(1): 37–62.

Kay, A. C. 1991. "Computers, Networks and Education." *Scientific American* 265, 3: 138–48.

Keller, Evelyn Fox. 1983. *A Feeling for the Organism: The Life and Work of Barbara McClintock.* San Francisco: W. H. Freeman.

Keller–Cohen, D. 1987. "Literate Practices in a Modern Credit Union." *Language in Society* 16: 7–23.

Kelley, D.A. 1983. *Documenting computer application systems: Concepts and Techniques.* New York: Petrocelli Books.

Keyes, E., D. Sykes, and E. Lewis. 1988. "Technology + Research + Design = Information Design. *IEEE Transactions on Professional Communications* PC-30: 32–37.

Killingsworth, Jimmie, and Michael Gilbertson. 1988. "How Can Text and Graphics be Integrated Effectively?" In Lynn Beene and Peter White, eds. *Solving Problems in Technical Writing.* New York: Oxford University Press.

Kinneavy, J. L. 1971. *A Theory of Discourse.* New York: W.W. Norton and Co.

Klein, Julie T. 1990. *Interdisciplinarity.* Detroit: Wayne State University Press.

Kostelnick, C. 1989. "Visual Rhetoric: A Reader-Oriented Approach to Graphics and Design." *The Technical Writing Teacher* XVI (1): 77–88.

Kraft, J. A. 1961. "A 1961 Compilation and Brief History of Human Factors Research in Business and Industry." *Human Factors* 3(4): 253–65.

Kranzberg, Marvin. 1967. "The Unity of Science-Technology." *American Scientist* 55(1): 48–66.

Krull, R. 1989. "If Icon, Why Can't You?" In S. Doheny-Farina, ed. *Effective Documentation: What We Have Learned From Research* (pp. 255–73). Cambridge, Mass.: MIT Press.

Kuhn, Thomas. 1970. *The Structure of Scientific Revolutions.* 2d edition. Chicago: The University of Chicago Press.

Lakoff, Robin. 1990. *Talking Power.* New York: Basic Books.

Landauer, Thomas K. 1995. *The Trouble with Computers: Usefulness, Usability, and Productivity.* Cambridge, Mass.: MIT Press.

Landow, George P. 1992. *Hypertext: The Convergence of Contemporary Critical Theory and Technology.* Baltimore: Johns Hopkins University Press.

Lannon, John. 1997. *Technical Writing.* 7th edition. New York: Longman Publishers.

Latour, Bruno. 1992. "Where Are the Missing Masses? The Sociology of a Few Mundane Artifacts." *Shaping Technology/Building Society: Studies in Sociotechnical Change.* Cambridge, Mass.: MIT Press.

Latour, Bruno, and Steve Woolgar. 1986. *Laboratory Life: The Construction of Scientific Facts.* Princeton: Princeton University Press.

Lauer, J. M. 1984. "Composition Studies: Dappled Discipline." *Rhetoric Review* 3 (1): 20–29.

Lauer, J. M. and J. W. Asher. 1988. *Composition Research: Empirical Designs.* New York: Oxford.

Lauer, J., and J. Atwill. 1995. "Refiguring Rhetoric: Aristotle's Concept of Techne." In Rosalind Gabin, ed. *Discourse Studies in Honor of James Kinneavy* (pp. 25–40). Matyland: Scripta Humanistica.

Law, John, ed. 1991. *A Sociology of Monsters: Essays on Power, Technology, and Domination.* London: Routledge Press.

Layton, Edwin. 1971. "Mirror-Image Twins: The Communities of Science and Technology in 19th Century America." *Technology and Culture* 12(3): 562–80.

———. 1976. "American Ideologies of Science and Engineering." *Technology and Culture* 17(3): 688–700.

———. 1974. "Technology as Knowledge." *Technology and Culture.* Vol. 15, No. 1. pp. 31-41.

———. 1971. *The Revolt of the Engineers: Social Responsibility and the American Engineering Profession.* Cleveland: Case Western University Press.

Levy, Steven. 1991. "Cyberspaced: A Gentle Demurral to the Hype About Virtual Reality." *Macworld* 8, 11: 61–68.

———. 1995. "The Luddites Are Back." *Newsweek* (June 12): p. 55.

Lewis, C. 1986. "Understanding What's Happening in System Interactions." In *User-Centered System Design: New Perspectives on Human-Computer Interaction,* (pp. 172–85). Hillsdale, N.J.: Lawrence Earlbaum Press.

Lewis, E. 1989. "Design Principles for Pictorial Information." In *Effective Documentation: What Research Tells Us.* (pp. 235–53). Cambridge, Mass.: MIT Press.

Long, Pamela. 1991. "The Openness of Knowledge: An Ideal in its Context in 16th-Century Writings on Mining and Metallurgy." *Technology and Culture.* 32 (2): pp. 318–55.

Low, Roderick. 1994. *Writing User Documentation: A Practical Guide For Those Who Want to Read.* New York: Prentice Hall.

Mackenzie, Donald. 1990. *Inventing Accuracy: A Historical Sociology of Nuclear Missle Guidance Systems.* Cambridge, Mass.: MIT Press.

Mackenzie, Donald and Wajcman, Judy. 1985. *The Social Shaping of Technology.* Milton Keynes: Open University Press.

Mathes, J. C. and D. W. Stevenson. 1976. *Designing Technical Reports.* Indianapolis: Bobbs-Merrill.

Mckay, L. 1984. *Soft Words, Hard Words: A Common Sense Guide to Creative Documentation.* Culver City, Calif.: Ashton-Tate.

Mikelonis, V. M. and V. Gervikas. 1985. "Using Computers in the Technical Writing Classroom: A Selected Bibliography 1978–1984." *The Technical Writing Teacher* 12(2): 161–76.

Mill, Gordon H., and Johan A. Walter. 1978. *Technical Writing.* 4th edition. New York: Holt, Rinehart, and Winston.

Miller, Carolyn. 1993. "Rhetoric and Community: The Problem of the One and the Many." In Theresa Enos and Stuart C. Brown, eds. *Defining the New Rhetorics.* Newbury Park: Sage Press.

———. 1989. "What's Practical About Technical Writing?" In B. E. Fearing, and W. K. Sparrow, eds., *Technical Writing: Theory and Practice* (pp. 14–24). New York: MLA.

———. 1979. "A Humanistic Rationale for Technical Writing." *College English* 40(6): 610–17.

Miller, G. A. 1956. "The Magical Number Seven Plus or Minus Two: Some Limits On Our Capacity For Processing Information." *Psychological Review* 63: 81–97.

Mirel, Barbara. 1988. "The Politics of Usability: The Organizational Functions of an In-House Manual." In S. Doheny–Farina, ed. *Effective Documentation: What We Have Learned From Research* (pp. 277–97). Cambridge, Mass.: MIT Press.

Mitcham, Carl. 1994. *Thinking Through Technology: The Path Between Engineering and Philosophy.* Chicago: The University of Chicago Press.

Moll, T., and R. Sauter. 1987. "Do People Really use On-Line Assistance?" In *Proceedings of Interact '87.* Amsterdam: North-Holland: 191–94.

Monk, A. 1984. *Fundamentals of Human–Computer Interaction.* Orlando: Academic Press.

Montaigne. 1952. "On Repentance." Chicago: Encyclopedia Britannica, Inc.

Moran, Michael G. and Debra Journet, eds. 1985. *Research in Technical Communication: a Bibliographic Sourcebook.* Westport, Conn.: Greenwood Press.

Moulthrop, Stuart. 1989. "In the Zones: Hypertext and the Politics of Interpretation." *Writing on the Edge* 1(1): 18–27.

Mumford, Lewis. 1934. *Technics and Civilization.* New York: Harcourt, Brace, & World.

———. 1964. "Authoritarian and Democratic Technics." *Technology and Culture* 5(1): 1–8

Myers, Greg. 1990. *Writing Biology: Texts in the Social Construction of Scientific Knowledge.* Madison, Wis.: Wisconsin University Press.

Nelson, Ted. 1987. *Dream Machines.* Redmond, Wash.: Tempus Books.

Newell, Allen and Stuart Card. 1986. "Straightening Out Softening Up: Response to Carroll and Campbell." *Human-Computer Interaction* 2: 251–67.

———. 1985. The Prospects for Psychological Science in Human–Computer Interaction. *Human–Computer Interaction* 1: 209–42.

Newell, A., and H. A. Simon. 1972. *Human Problem Solving.* Englewood Cliffs, N.J.: Prentice Hall.

Nielsen, Jakob. 1993. *Hypertext and Hypermedia.* Boston, Mass.: Academic Press.

Norman, D. 1987. "Cognitive Science-Cognitive Engineering." In J. M. Carroll, ed. *Interfacing Thought* (pp. 325–65). Cambridge, Mass.: MIT Press.

Norman, Donald. 1988. *The Design of Everyday Things.* New York: Basic Books.

Norman, D., and S. Draper, 1986. *User-Centered System Design: New Perspectives on Human–Computer Interaction.* Hillsdale, N.J.: Lawrence Earlbaum Press.

Oborne, D. J. 1989. "Ergonomics and Information Technology." In T. Forester, ed. *Computers in the Human Context: Information Technology, Productivity, and People* (pp. 174–87). Cambridge, Mass.: MIT Press.

Odell, L., and D. Goswami. 1985. *Writing in Nonacademic Settings.* New York: Guilford Press.

Ong, Walter J. 1971. *Rhetoric, Romance and Technology*. Ithaca, N.Y.: Cornell University Press.

———. 1975. "The Writer's Audience Is Always a Fiction." *PMLA* 90: 9–21.

Palmer, J., and T. Duffy, et. al. 1988. "The Design and Evaluation of On-line Help for UNIX EMACS: Capturing the User in Menu Design." *IEEE Transactions on Professional Communications* 31 (1): 44–51.

Papert, S. 1980. *Mindstorms: Children, Computers and Powerful Ideas*. New York: Basic Books.

Paradis, James. 1991. "Text and Action: The Operator's Manual in Context and in Court." In Charles Bazerman and James Paradis, eds. *Textual Dynamics of the Professions: Historical and Contemporary Studies of Writing in Professional Communities*. Madison, Wis.: Wisconsin University Press.

Park, D. 1982. "The Meanings of Audience." *College English* 44: 247–57.

———. 1986. "Analyzing Audiences." *College Composition and Communication* 37 (4): 478–86.

Phelps, L. W. 1988. *Composition As a Human Science: Contributions to the Self-Understanding of a Discipline*. New York: Oxford University Press.

Pinch, T. J. and W. E. Bijker. 1987. "The Social Construction of Facts and Artifacts: Or How the Sociology of Science and the Sociology of Technology Might Benefit Each Other." In Bijker et. al., eds. *The Social Construction of Technological Systems: New Directions in the Sociology and History of Technology* (pp. 17–49). Cambridge, Mass.: MIT Press.

Plato. 1952. *The Dialogues of Plato*. Translated by Benjamin Jowett. Chicago: *Encyclopedia Britannica*, Inc.

Polya, G. 1945. *How to Solve It*. Princeton: Princeton University Press.

Polyani, Michael. 1958. *Personal Knowledge: Toward a Post–Critical Philosophy*. Chicago: The University of Chicago Press.

Porter, J. 1992. *Audience and Rhetoric: An Archeological Composition of the Discourse Community*. Englewood Cliffs, N.J.: Prentice Hall.

Powell, Malea. 1994. "This Is How I Heard It: Stories Toward an American Indian Theorhetoric." Unpublished master's thesis, Miami University: Oxford, Ohio.

Price, J. 1984. *How To Write a Computer Manual: A Handbook of Software Documentation*. Menlo Park, Calif.: The Benjamin Cummings Publishing Company.

———. 1988. "Creating a Style for On-line Help." In E. Barrett, ed. *Text, Context, Hypertext: Writing With and For the Computer* (pp. 329–42). Cambridge, Mass.: MIT Press.

Price, Jonathan, and Henry Korman. 1993. *How to Communicate Technical Information: A Handbook of Software and Hardware Documentation.* Redwood City, Calif.: The Benjamin Cummings Publishing Co.

Ramey, J. 1986. "Developing a Theoretical Base for On-line Documentation Part I: Building the Theory." *The Technical Writing Teacher* 13 (2): 148–59.

———. 1986. "Developing a Theoretical Base for On-line Documentation Part II: Applying the Theory." *The Technical Writing Teacher* 13 (3): 302–15.

———. 1989. "How People Use Computer Documentation: Implications For Book Design." In S. Doheny-Farina, ed. *Effective Documentation: What We Have Learned From Research* (pp. 143–57).Cambridge, Mass.: MIT Press.

Redish, J. C. 1988. "Reading To Learn To Do." *The Technical Writing Teacher* 15: 223–33.

Redish, J. C., R.M. Battison, and E.S. Gold. 1985. "Making Information Accessible to Readers." In L. Odell and D. Goswami, eds. *Writing in Nonacademic Settings* (pp. 29–153). New York: The Guilford Press.

Reichert, Pegeen. 1996. "A Contributing Listener and Other Composition Wives: Reading and Writing the Feminine Metaphors in Composition Studies." *The Journal of Advanced Composition.* 16 (1): 141–57.

Reitman, P. 1988. "Streamlining Your Documentation Using Quick References." *IEEE Transactions on Professional Communications* 31: 75–83.

Richards, I. A., and C. K. Ogden. 1923. "The Meaning of Meaning." In P. Bizzell and B. Herzberg, eds. *The Rhetorical Tradition: Readings From Classical Times to the Present* (pp. 967–74). New York: St. Martin's Press.

Ridgway, L. 1987. "Read My Mind: What Users Want From On-line Information." *IEEE Transactions on Professional Communications* PC-30, 2: 87–90.

Rodriguez, D. 1985. "Computers and Basic Writers." *College Composition and Communication* 36: 336–39.

Rubens, B. 1989. "Similarities and Differences in Developing Effective On-line and Hardcopy Tutorial." In eds. *Effective Documentation: What We Have Learned From Research* (pp. 157–83). Cambridge, Mass.: MIT Press.

Rubens, P., and R. Krull. 1988. "Designing On-line Information." In *Text, Context, Hypertext: Writing With and For the Computer* (pp. 291–310). Cambridge, Mass.: MIT Press.

Rubens, P., and B. Rubens. 1989. "Usability and Format Design." In *Effective Documentation: What We Have Learned From Research* (pp. 213–23). Cambridge, Mass.: MIT Press.

Rutter, Russell. 1991. "History, Rhetoric, and Humanism: Toward a More Comprehensive Definition of Technical Communication." *Journal of Technical Writing and Communication* 21(2): 133–53.

Scarfe, Francis, ed. 1964. *Baudelaire: Selected Verse.* Middlesex, U.K.: Penguin Press.

Schneiderman, B. 1987. *Designing the User Interface: Strategies for Effective Human-Computer Interaction.* Reading, Mass.: Addison-Wessley.

Schneiderman, Ben. 1992. "Socially Responsible Computing I & II: A Call to Action Following the L.A. Riots." *SIGCHI Bulletin* (July 1992): 14–17.

Sedgwick, John. 1993. "The Complexity Problem." *Atlantic Monthly* (March): 96–104.

Seiden, P., and P. Sullivan. 1986. "Designing User Manuals for the On-line Public Access Catalog." *Library HiTech* 4 (13): 29–36.

Shirk, H. N. 1988. "Technical Writers as Computer Scientists: The Challenges of On-line Documentation." In E. Barrett, ed. *Text, Context, Hypertext: Writing With and For the Computer* (pp. 311–28). Cambridge, Mass.: MIT Press.

Silvers, Robert B., ed. 1995. *Hidden Histories of Science.* New York: *New York Review of Books.*

Simon, Herbert A. 1970. *The Sciences of the Artificial.* Cambridge, Mass.: MIT Press.

———. 1977. *The New Science of Management Decision.* Englewood Cliffs, N.J.: Prentice Hall.

Simpson, Mark. 1989. *Shaping Computer Documentation For Multiple Audiences: An Ethnographic Study.* Unpublished dissertation. W. Lafayette, Ind.: Purdue University.

Singer, Charles. 1956, 1957. *The History of Technology. 2 vol.* Oxford: Oxford University Press.

Slatin, John M. 1990. "Reading Hypertext: Order and Coherence in a New Medium." *College English* 52(8): 870–83.

Smith, D. B. 1986. "Axioms for English in a Technical Age." *College English* 48(6): 567–79.

Smith, Merritt R., and Leo Marx. 1994. *Does Technology Drive History?: The Dilemma of Technological Determinism.* Cambridge, Mass.: MIT Press.

Spenser, Edmund. 1977. *The Faerie Queene.* A. C. Hamilton, ed., 3d ed. London, New York: Longman.

Spilka, R. 1988. "Studying Writer–Reader Interactions in the Workplace." *The Technical Writing Teacher* 15(3): 208–21.

Star, Susan Leigh. 1995. "Epilogue: Work and Practice in Social Studies of Science, Medicine, and Technology." *Science, Technology, and Human Values.* 20 (4): 501–07.

———. 1991. "Power, Technologies and the Phenomonology of Conventions: On Being Allergic to Onions." In John Law, ed., *A Sociology of Monsters: Essays on Power, Technology, and Domination* (pp. 26–56). London: Routledge.

Staudenmaier, John. 1985. *Technology's Storytellers.* Cambridge, Mass.: MIT Press.

Sticht, T. 1985. "Understanding Readers and Their Uses of Text." In T. M. Duffy and R. Waller, eds. *Designing Usable Text* (pp. 315–40). Orlando, Fla.: Academic Press.

Sticht, T. G. 1977. "Comprehending Reading at Work." In M. A. Just and P. A. Carpenter, eds. *Cognitive Processes in Comprehension.* Hillsdale, N.J.: Erlbaum.

Suchman, Lucy. 1987. *Plans and Situated Actions: The Problem of Human–Machine Communication.* Cambridge, England: Cambridge University Press.

———. 1994. "Working Relations of Technology Production and Use." *Computer Supported Cooperative Work* 2: 21–39.

Sullivan, Dale. 1990. "Political–Ethical Implications of Defining Technical Communication as a Practice." *Journal of Advanced Composition* 10(2): 375–86.

Sullivan, Patricia. 1989. "Beyond a Narrow Conception of Usability Testing." *IEEE Transactions on Professional Communication.* 32 (4): 256–64.

———. 1988. "Writers as Total Desktop Publishers: Developing a Conceptual Approach to Training." In E. Barrett, ed. *Text, Context, Hypertext: Writing With and For the Computer* (pp. 265–78). Cambridge, Mass.: MIT Press.

Sullivan P., and L. Flower. 1986. "How Do Users Read Computer Manuals?: Some Protocol Contributions to Writers' Knowledge." In B. T. Peterson, ed. *Transactions* (pp. 163–78). Urbana, Ill.: NCTE.

Sullivan, P., and P. Seiden. 1985. "Educating On-line Catalog Users: The Protocol Assessment of Needs." *Library HiTech* 3(10): 11–19.

Sullivan, Patricia and James E. Porter. 1993. "Remapping Curricular Geography: Professional Writing In/And English." *Journal of Business and Technical Communication* 7 (4): 389–422.

———. 1992. "On Theory, Practice and Method: Toward a Heuristic Research Methodology for Professional Writing." In R. Spilka, ed. *Research Perspectives on Writing in the Workplace.* Carbondale, Ill.: SIU Press.

Tebeaux, Elizabeth, and M. Jimmie Killingsworth. 1993. "Expanding and Redirecting Historical Research in Technical Writing: In Search of Our Past." *Technical Communication Quarterly* 1 (2): 5–32.

Tinker, Miles Albert. 1965. *Bases for Effective Reading.* Minneapolis, Minn.: University of Minnesota Press.

Trimbur, J. 1989. "Consensus and Difference in Collaborative Learning." *College English* 51: 602–16.

Tufte, E. R. 1983. *The Visual Display of Quantitative Information.* Cheshire, Conn.: Graphics Press.

Tuman, Myron C. 1992. *Wordperfect: Literacy in the Computer Age.* Pittsburgh: Pittsburgh University Press.

Vernant, Jean–Pierre. 1982. *The Origins of Greek Thought.* Ithaca: Cornell University Press.

Wajcman, Judith. 1991. *Feminism Confronts Technology.* University Park, Penn.: Pennsylvania State University Press.

Weiss, E. H. 1991. *How To Write Usable User Documentation.* Phoenix: Oryx Press.

Wells, Susan. 1986. "Jurgen Habermas, Communicative Competence, and the Teaching of Technical Discourse." In Cary Nelson, ed. *Theory in the Classroom* (pp. 245–69). Urbana, Ill.: Illinois University Press.

White Jr., Lynn. 1978. *Medieval Religion and Technology: Collected Essays.* Berkeley: California University Press.

Wieber, Norbert. 1948. *Cybernetics.* New York: John Wiley & Sons.

———. 1950. *The Human Use of Human Beings: Cybernetics and Society.* New York: Houghton Mifflin.

Williams, Mark. V., et al. 1995. "Inadequate Functional Health Literacy Among Patients at Two Hospitals." *Journal of the American Medical Association* 274 (21): 1677–82.

Winner, Langdon. 1977. *Autonomous Technology: Technics Out-Of-Control as a Theme in Political Thought.* Cambridge, Mass.: MIT Press.

———. 1985. *The Whale and the Reactor.* Chicago: The University of Chicago Press.

———. 1993. "Upon Opening the Black Box and Finding it Empty: Social Constructivism and the Philosophy of Technology." *Science, Technology, and Human Values* 18(3): 362–78.

Winograd, T., and F. Flores. 1986. *Understanding Computers and Cognition: A New Foundation for Design.* Norwood, N.J.: Ablex.

Winsor, Dorothy. 1990. "The Construction of Knowledge in Organizations: Asking the Right Questions About the Challenger." *Journal of Business and Technical Communication* 4: 7–21.

Woolgar, Steve. 1991. "The Turn to Technology in Social Studies of Science." *Science Technology and Human Values* 161: 20–50.

Wright, Patricia. 1988. "The Need For Theories of NOT Reading: Some Psychological Aspects of the Human–Computer Interface." In Ben A. G. Elsendoorn and Herman Bouma, eds. *Working Models of Human Perception* (pp. 319–40). London: Academic Press.

———. 1983. "Manual Dexterity: A User-Oriented Approach to Creating Computer Documentation." *CHI '83 Proceedings* (December): 11–18.

Yates, JoAnne. 1989. *Control Through Communication: The Rise of System American Management.* Baltimore, Md.: Johns Hopkins University Press.

Young, Richard E. 1980. "Arts, Crafts, Gifts and Knacks: Some Disharmonies in the New Rhetoric." *Visible Language* 14(4): 341–50.

Index

37101097R00123

Made in the USA
Lexington, KY
18 November 2014